Geology and Radwaste

Academic Press Geology Series

Mineral Deposits and Global Tectonic Settings—A. H. G. Mitchell and
M. S. Garson—1981

Applied Environmental Geochemistry—I. Thornton (ed.)—1983

Geology and Radwaste—A. G. Milnes—1985

Geology and Radwaste

A. G. MILNES

Department of Earth Sciences
Swiss Federal Institute of Technology
Zürich, Switzerland

1985

ACADEMIC PRESS

(Harcourt Brace Jovanovich, Publishers)

London   Orlando   San Diego   New York
Toronto   Montreal   Sydney   Tokyo

ACADEMIC PRESS INC. (LONDON) LTD.
24-28 Oval Road
LONDON NW1 7DX

United States Edition published by
ACADEMIC PRESS, INC.
Orlando, Florida 32887

BRITISH LIBRARY CATALOGUING IN PUBLICATION DATA
Milnes, A.G.
 Geology and radwaste. — (Geology series)
 1. Radioactive waste disposal
 I. Title   II. Series
 621.48′38′01551   TD812

LIBRARY OF CONGRESS CATALOGING-IN-PUBLICATION DATA
Milnes, A. G. (Arthur George), Date
 Geology and radwaste.

 Includes bibliographies and index.
 1. Geology.   2. Environmental protection.   3. Radio-
active waste disposal.   I. Title.
QE33.M475   1985   621.48′38   85-1229
ISBN 0-12-498070-8 (alk. paper)
ISBN 0-12-498071-6 (paperback)

PRINTED IN THE UNITED STATES OF AMERICA

85 86 87 88   9 8 7 6 5 4 3 2 1

To Kirsten

"For all that has been—Thanks!
For all that shall be—Yes!"

Dag Hammarskjöld

CONTENTS

PART II. EARTH SCIENCE PERSPECTIVES

3. The Earth's Crust

4. Geological Time

11. Fluid–Rock Interaction

12. Ocean Processes

13. Climatic Change and Continental Glaciation

PART III. APPLICATION

14. Predictive Geoscience

15. Repository Site Selection

PREFACE

One of the most prominent controversies in the public debate on the future of nuclear energy concerns the safe disposal of the radioactive wastes derived from uranium mining and milling, from nuclear fuel fabrication and reprocessing, and from the running of nuclear reactors. Whether or not the prominence attached to this problem is justified, it has focused attention on the earth sciences in a unique way. Scientists, politicians and the public at large are demanding answers to questions which have hardly been asked before, or only been accorded passing academic interest. The inability of geologists to provide quick, assured solutions, and the patent lack of consensus among earth scientists on many points, has stimulated interest in wide circles, a desire to know more about how the Earth works, why this knowledge is important for the problem at hand, and why so many uncertainties still remain. This book is written with this wider audience in mind. The first part is general in scope, outlining the magnitude and nature of the radwaste disposal problem and the types of solutions which have been proposed. The second part is specifically geological, and the level of detail included is aimed at geology students in the later stages of their studies, as well as at any professional geologist who is not directly involved in radwaste disposal research. However, a non-technical style and a simple structure have been chosen in an attempt to make it comprehensible to a much broader group, in fact, to anyone with some scientific education and an interest in environmental problems. After two introductory chapters providing geological background material, Part II quickly becomes highly selective, following those lines of enquiry most relevant to the long-term safety of the different disposal systems. A review of the topics selected, however, reveals the unique nature of the radwaste problem: in spite of the selectivity, it touches on so many aspects of geology that a broad overview of the earth sciences does finally emerge. The last part of the book is again of more general interest, illustrating the difficulties of actually applying geological knowledge and thinking to practical questions, such as risk assessment and site selection. It brings the book back "down to earth," back to the unfortunate reality of the involvement of science with politics, which has been the hallmark of the radwaste problem from the beginning.

This book is a personal digest of the enormous volume of literature which has appeared on the subject of geology and radioactive wastes in recent years (up to and including 1983). As an academic, I have not been directly involved in radwaste disposal research, except in an advisory capacity. My initial interest in the subject was prompted by a growing concern about the destructive tendencies of modern technology, and about the increasingly alarming effects these tendencies are already having on the quality of human life and the natural environment. Among other things, this has led me to a conviction which I still hold, that is, that the "nuclear road," emphasizing the satisfaction of our ever-growing energy "needs" with ever-increasing numbers of nuclear power stations, and leading to a state of unlimited energy availability based on fast breeder and fusion technology, is *not* the road that society should be taking. Paradoxically, however, I found that my assessment of the radioactive waste problem neither strengthened nor weakened this conviction. It became clear to me that radioactive wastes constitute an environmental hazard which in no way overshadows the problems associated with many other waste products (CO_2, SO_2, heat, chemicals, heavy metals, etc.) whose management is haphazard, to say the least, and whose "disposal" has not yet led to widespread public concern. The positive side of the historical accident which has fixed such enormous attention on what may, in retrospect, seem to be an insignificant part of the whole process of environmental degradation is that it is laying a solid foundation for the better management of all other types of hazardous waste. For me personally, the negative side is that such a great effort should have to be put into alleviating the side-effects of an activity which I believe to be unnecessary and misguided in the first place. Nevertheless, I recognize that there is no turning back the clocks. I am convinced that, now, even a turning away from the nuclear road requires intensive work toward achieving satisfactory solutions to the problems of radwaste disposal. This is well illustrated by the situation in Sweden, where the proposed phasing out of nuclear energy by the year 2010 still leaves the next generation with a serious environmental problem, unless present efforts at solving it are completely successful. Recognition of this fact provided the main motivation for the writing of the present book.

Many people have contributed, through lively discussion and supportive interest, to the final product. For this, I wish to thank them warmly, and, at the same time, to relieve them of any responsibility for what eventually was included or left out, or what opinions are expressed. I am particularly grateful to M. Buser, E. Frank, K. Kelts, J. G. Ramsay, G. Skippen, R. Trümpy, H. Weissert and W. Wildi for reading and criticizing early drafts of the text and for all their helpful comments. P. Huggenberger and G. Juvalta provided invaluable help in searching and keeping up with the literature, and U. Gerber and A. Uhr, in producing the text figures. The final manuscript was immaculately typed by Audrey Haas. My wife and children sometimes aggravated, sometimes ameliorated the inevitable crises, but their never-failing sympathy saw me through.

PART I

BACKGROUND

The disposal of radioactive wastes has become a major environmental problem through the widespread use of nuclear fission reactors for the manufacture of nuclear weapons and for the production of electricity. The quantities and types of radwaste are reviewed in Chapter 1 with reference to the nuclear fuel chain, the sequence of processes which starts with the mining of uranium ore and ends with the removal of "burnt-out" nuclear fuel from the reactor core after use. These are compared with the radwastes expected from proposed nuclear fuel cycles, in which the spent fuel is reprocessed and returned to the reactor, and other future sources. Although quantities and types vary, for the purposes of discussing the geological aspects of radwaste disposal, two categories can be distinguished, low-level and high-level, according to the long-term health hazard. Present and proposed methods of disposal of radwastes in these two categories are summarized in Chapter 2. These are grouped into sections on past and present practices for liquid and solid low-level waste disposal (including deep well injection, grout injection, shallow land burial, and ocean dumping) and proposed concepts for solid high-level waste disposal (such as immobilization, deep mine or borehole emplacement, deep underground melting, and ice sheet and subseabed emplacement). This serves as a background for Part II, in which the long-term effects of natural processes on the various types of radwaste isolation system are discussed from the perspective of the earth sciences.

CHAPTER 1

Radioactive Wastes

I. INTRODUCTION

Wastes are unwanted, useless and left-over materials, movable materials owned by someone who wishes to get rid of them. Wastes need to be disposed of in a regulated way for the general good. Since the production of inordinate amounts of waste is a characteristic of industrial society, waste treatment, waste management and waste disposal have become unattractive but indispensable activities involving a significant part of the total work and investment effort. Apart from the problem of the ever-increasing quantities, some waste materials represent an immediate or a long-term hazard to life or health in whatever quantity they may occur. Of all the types of hazardous waste known, that which has most often been the focus for public concern in recent years is waste which emanates ionizing radiation. Radioactive waste has become a kind of symbol for the many people who question the assumption that the benefits of industrialization always outweigh its detrimental effects. Yet the experts are unanimous: radioactive wastes, properly treated and managed, present a far smaller danger than many other waste materials for which society in general shows little or no concern (Fig. 1). We take this contrast between hazard as perceived by the scientist and danger as perceived by the general public as the starting point of this brief introductory review of the sources and properties of the different types of radwaste.

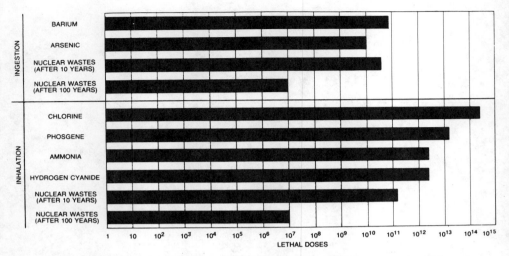

Fig. 1. One way of illustrating the risks associated with radioactive wastes, in relation to risks associated with other activities: comparison of health hazards presented by high-level radioactive wastes with those of other poisonous substances routinely in use. Comparison is based on the total number of lethal doses of different materials known to exist in the United States at the present time. Because nuclear wastes are closely guarded, whereas the other substances are "routinely scattered around on the ground where food is grown," this comparison grossly overestimates the actual radwaste hazard. (From Cohen, 1977, by permission of W. H. Freeman and Co.)

II. SOURCES OF RADWASTE

The use of radioactive materials is today a major facet of industrial society. They are important tools for diagnosis in medicine; they provide power for such life-saving devices as cardiac pacemakers and heart pumps; they drive submarines and ships; they support many of the small energy sources required for navigation, weather forecasting and communications at sea, in the air and in space; and they are used as sources for process radiation in the chemical, petrochemical and food industries. All these usages are increasingly widespread and the benefits accruing are generally considered far to outweigh the risks involved. However, there has still been little discussion of the radioactive waste problem associated with these activities. In fact, one report blithely illustrates how the plutonium power unit of a satellite reenters the atmosphere after use, plunges into the ocean and sinks to the bottom with the terse heading "Impact and ultimate disposal!"

The radioisotopes (Section III,B) used in research, medicine and industry are mainly produced by neutron bombardment of nonradioactive materials in nuclear reactors, although some are naturally occurring (e.g., radium) and some are obtained by bombardment with protons or other charged particles in an accelerator (e.g., beryllium-7). Combining bombardment with chemical or physical separation procedures enables radioisotopes of practically every element to be manufactured, and these processes themselves are a further source of radwaste. However, both the activities mentioned above—the use and the manufacture of radioisotopes—yield only minor amounts of radioactive waste in comparison to the nuclear reactors themselves and the preparation of reactor fuel.

Nuclear reactors are designed to sustain a controlled fission process in which the fuel (generally the naturally occurring radioisotope uranium-235) is split into two lighter elements by neutron bombardment, thus releasing further neutrons to continue the process. This chain reaction, once established, releases large amounts of energy which are generally used for electricity production. The process itself also produces a whole range of radioisotopes, including plutonium-239, which is the raw material for the manufacture of nuclear weapons. It is the production and use of reactor fuel which is the main source of radioactive waste, so these activities will be taken as a basis for the following discussion.

A. The Nuclear Fuel Chain ("Once-Through" Option)

We shall approach the question of the size of the radwaste problem by looking more closely at the nuclear fuel chain: the sequence of steps which lead from the mining of the raw materials, through the manufacture of the fuel and the "burning" of it in the reactor, to its removal from the reactor once it is "burnt out" or spent (Fig. 2). In this section, we will concentrate on the amounts of radwaste produced at each stage and for this purpose we shall use a rough subdivision into wastes of low radioactivity and wastes of high radioactivity, according to whether they can be handled immediately without shielding and with a minimum of remote control, or whether they require thick shielding and completely remote handling and must be stored and cooled for long periods of time before transport. In the following section on waste properties, we shall see that this subdivision is unsatisfactory from the point of view of long-term waste management and we will go on to a more relevant classification.

Nuclear fuel is made out of uranium ore. The ore is found in many different geological environments, but usually as diffuse accumulations in sedimentary rocks, in concentrations of 0.15–0.30% of uranium oxide U_3O_8. A total of 100,000 tons or more of ore have to be mined, with much larger

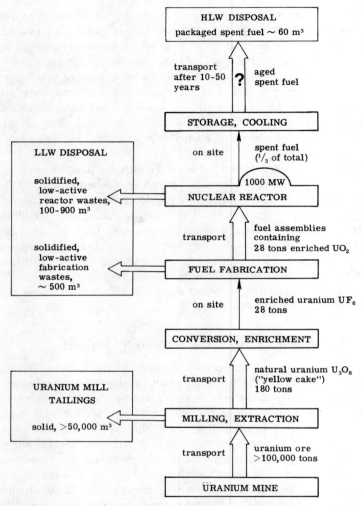

Fig. 2. Flow sheet for nuclear fuel and radwaste along the nuclear fuel chain (the "once-through" option). The fuel flow represents that necessary to run a 1000 MW nuclear reactor for 1 year. The waste quantities should be taken as order-of-magnitude only. Radioactive gaseous effluents have been omitted. (Compiled from Organization of Economic Cooperation and Development, 1977; Hermann, 1983; Carter *et al.*, 1979 and SIPRI, 1979.)

amounts of barren rock and subgrade material, to provide the fuel for a single reactor (1000 MW) for 1 year, and since the most concentrated ore bodies are discovered and exploited first, these quantities are going to increase rapidly in the future. However, the wastes produced by uranium mining have such low levels of radioactivity, similar to those of many common rocks, that they are not usually classified as radwaste. The environmental problems which arise from increased demands for uranium ore will be governed by the huge quantities of waste and the large amounts of energy necessary for processing, not by the release of radioactivity.

The uranium ore is transported to a uranium mill as the first step in processing. There it is pulverized (grain sizes 0.001–1 mm, according to ore type) and transferred to either acid or alkali leach tanks (according to host rock type) to dissolve out the uranium. The dissolved uranium is purified, concentrated and solidified by a variety of processes which lead to the basic end product, natural uranium, or yellow cake, containing 75–85% U_3O_8. Wastes of low radioactivity are produced in very large quantities during the milling processes. First, the large amounts of liquid used in leaching (about 2000 litres/ton of ore) contain, in addition to aggressive chemicals, appreciable amounts of radioisotopes such as radium-226 and thorium-230 as well as 10% of the original uranium. Second, the sludges remaining after leaching (the mill tailings) are contaminated with the same radionuclides and exceed in volume the original ore. The sludges and waste liquids are generally combined in large tailings ponds, which dry out by seepage and evaporation. Each ton of yellow cake produced leaves > 300 m^3 of solid low-active waste to be disposed of, a major environmental problem to be addressed later in the book (see Chapter 2, Section III,D and Chapter 5, Section V).

The yellow cake is then transported to the industrial complex at which the fuel for the reactors is prepared. Here, there are three main steps: conversion, enrichment, and fuel fabrication (Fig. 2). Enrichment entails increasing the concentration of the uranium isotope ^{235}U, the actual fissionable material, from its natural value of 0.7% to $\sim 3\%$ of the total uranium content, which is mainly ^{238}U. This can be done by various procedures, each requiring the conversion of the yellow cake into a corresponding form. The most common process is to convert it to the volatile fluoride, UF_6, and then to pass the gas through thousands of stages containing diffusion barriers which cause the isotopes to be separated on the basis of their different atomic masses. This produces two gas streams, one depleted and one enriched in ^{235}U. Enrichment to 3% produces about five times as much depleted as enriched uranium; the depleted material, however, is not treated as waste but stored for possible future use (e.g., in future thorium-fueled or fast breeder reactors). The enriched uranium goes on to the fuel fabrication unit, where it is converted into solid uranium oxide, UO_2, compressed and formed into pellets. These are loaded into zirconium alloy tubes, which are then sealed and assembled into fixed arrays, or fuel assemblies, before being transported to the reactor. The wastes resulting from conversion, enrichment and fuel fabrication are all of low radioactivity and are minor in quantity. For each ton of natural uranium processed, about 20,000 litres of liquid waste and 2 m^3 of solid waste are produced, as well as some low-active gaseous and particulate effluents.

The radioactive wastes originating from the normal running of a conventional nuclear reactor of any type can be split into two groups: those produced during the everyday running of the reactor and those produced during the yearly close-down, when part of the spent fuel is removed from the reactor core and replaced by new fuel. The first group includes a wide range of low-active solids and liquids, such as contaminated clothing, machinery and components, ion exchange resins, evaporator concentrates, filter sludges and coolant liquids. The yearly volume (after solidification and packaging) varies from 100 to 900 m^3, depending on the reactor type, waste treatment and classification scheme. Some of these wastes require shielding and remote handling, but most do not. In contrast, the spent fuel assemblies (fuel rods and cladding hulls) removed from the reactor core are very highly radioactive and generate large amounts of heat. On a volume-to-volume basis, spent reactor fuel on removal is 10^8 times as radioactive as uranium mill tailings and 10^4–10^6 times as radioactive as the low-active reactor wastes. Removal, transport and emplacement in the interim storage pools next to the reactor must be carried out by remote control and the stored material must be continually cooled for several years until its self-heating has declined sufficiently to allow off-site transport.

Implicit in this brief summary of the nuclear fuel chain is that the spent fuel will ultimately be treated as waste material (see Fig. 2). This is the basic ingredient of what has become known as the "once-through" option for the peaceful use of atomic energy. Advocates of this procedure point out that it would considerably simplify the problem of radwaste disposal and that it would prevent the development of a commercial plutonium industry, with its attendant dangers of accidental releases to the environment, proliferation of nuclear weapons, and coupling of civil and military nuclear industries. At the moment of writing, the world stands at a significant crossroads in history. The spent fuel from its civilian reactors is lying in temporary storage awaiting a final decision. There is no doubt that the world would be a safer place to live in if the "once-through" option were chosen, but it is equally clear that uranium supply problems would then limit the widespread use of nuclear energy to the next century. Having large amounts of spent fuel at hand will probably, in the long run, constitute an irresistible temptation to use it as an energy source, particularly since, thanks to the nuclear weapons industry, all the required technology for recycling is well developed. For these reasons, the nuclear industry has for many years been moving towards the commercial reprocessing of spent nuclear fuel, rather than its final irretrievable disposal, and the consequences of this procedure for radwaste disposal will now be considered.

B. Nuclear Fuel Cycles (Reprocessing Options)

The reprocessing of spent reactor fuel has been carried out for many years in military installations to separate the plutonium needed for making nuclear weapons. Plutonium develops in the fuel during fission by the neutron bombardment of the uranium. The average amount of plutonium produced by a 1000-MW reactor in 1 year is about 270 kg (Fig. 3). Commercial reprocessing plants are modelled on their military predecessors, adapted for the needs of fuel recycling and for the desire to produce as little weapons-grade material as possible (antiproliferation measures). Pilot or small-scale commercial plants are at present operating in France (La Hague), the United Kingdom (Windscale), West Germany (Karlsruhe), India (Tarapur), and Japan (Tokai Mura), and numerous large plants are planned to be operating by the end of the decade, possibly including some of the mothballed U.S. commercial facilities. In spite of the long-standing and outspoken intentions of the nuclear industry to move towards fuel recycling, implementation has been very slow due to inherent technical difficulties (e.g., Windscale explosion, 1973), political actions (e.g., President Carter's antiproliferation measures, 1977), and escalating costs (resulting in price increases of 2000% or more over the past 15 years). Quite apart from these hindrances, however, there is still a lack of consensus on which of several possible nuclear fuel cycles is the most favourable, which in turn affects which of several reprocessing techniques should be used. It lies outside the scope of this book to review the consequences for the radwaste problem of all these, largely theoretical, alternatives. Rather, we look more closely at one of the more widely discussed options: the uranium–plutonium fuel cycle, suggested to produce mixed-oxide fuels for conventional light-water reactors, with a view to determining what its general introduction would imply.

For uranium and plutonium recycling (Fig. 3), a reprocessing plant is required to separate the U and Pu from the other radioisotopes in the spent fuel. This would be accomplished by a sequence of chemical and physical processes as follows:

1. Disassembly and chopping up of the spent fuel elements
2. Nitric acid leaching of the fuel from the zirconium alloy cladding hulls
3. Preparation of the leach solutions for solvent extraction
4. Solvent extraction to separate uranium and plutonium from the waste stream (containing the fission products and the non-Pu transuranics)
5. Purification of the uranium and plutonium streams, conversion to the oxides and preparation for transport to the enrichment or fuel fabrication plants
6. Solidification of the waste stream and immobilization in glass or a similar matrix

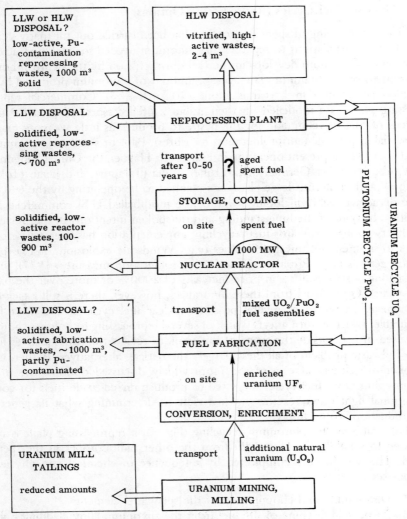

Fig. 3. Flow sheet for nuclear fuel and radwaste in one of the proposed nuclear fuel cycles (uranium/plutonium recycle). The fuel quantities are estimated to run a 1000 MW nuclear reactor for 1 year, based largely on models rather than actual practice. The waste quantities are correspondingly uncertain and are to be taken as rough order-of-magnitude estimates. Gaseous effluents are omitted. Note that the initial part of the nuclear fuel chain, uranium mining and milling, is still necessary for maintaining this cycle. (Compiled from American Physical Society, 1977; Cohen and King, 1978; Carter *et al.*, 1979 and Hermann, 1983.)

All these processes must be carried out by remote control, behind heavy shielding, and involve large amounts of material (equipment, components, tools, chemicals, process fluids, and filters) which eventually become contaminated with small quantities of radioisotopes. From the point of view of radwaste, then, one result of reprocessing will be to increase enormously the amount and variety of materials for disposal (compare Figs. 2 and 3).

The main problem with recycled wastes, however, is not the increased quantities (the amount of waste of high radioactivity is actually reduced). It is the fundamental change in character. A large proportion of the wastes of low radioactivity (those which can be handled directly, with only light shielding) are contaminated with small amounts of transuranic elements, like plutonium, which are long-lived and extremely toxic even in minute quantities. This applies not only to the radwastes from the reprocessing plant itself, but also to those of the mixed-oxide fuel fabrication facility. A yearly amount of 6–15 kg plutonium per 1000-MW reactor, which would be a total of 1000–2000 kg a year from the presently operating civil reactors of the world, would effectively be released to the biosphere, if such wastes were disposed of by present low-active waste disposal techniques (see Chapter 2). Because the hazard index of plutonium and other transuranic elements (see Section III,B) is so much higher than that of the fission products present in normal "once-through" reactor wastes of low radioactivity, it is clear that transuranic-contaminated radwastes will have to be treated much more carefully, a question we return to in the next section. Here, I simply wish to emphasize that reprocessing affects the radwaste problem profoundly in that it increases its complexity by widening the spectrum of radioactive materials which have to be dealt with and by introducing considerable uncertainty.

C. Future Sources of Radwaste

Several important sources of radioactive wastes are not included in the above estimates because amounts and characteristics can only be guessed at at the present time. A major group which will certainly have to be dealt with by future generations is that derived from the decontamination and decommissioning of shut-down nuclear reactors and reprocessing plants after their expected operating lifetime of 30–40 years. It will be necessary to mothball a shut-down reactor for many decades before the highly radioactive reactor vessel could be dismantled and when this becomes possible, a projected 50,000 m^3 of solid wastes with low activity per 1000-MW light water reactor will have to be disposed of. This aspect of radwaste management has only recently been subjected to detailed scrutiny and the cleanup and repair of the damaged Three Mile Island reactor indicates the enormous problems and costs involved.

A second future source of radioactive waste will be the commercial fast breeder reactors which are the logical future development if reprocessing, and not the "once-through" option, is chosen. Present systems using a liquid sodium coolant, for instance, produce considerable amounts of wastes of low and high activity which are not acceptable for disposal or reprocessing because of their content of elemental sodium. A further source of radwaste which may be of concern in the future is that associated with fusion reactors. Although difficult to estimate at present, the use of tritium (^{3}H) as a fuel and its production from Li during fusion processes produces a major hazard because of its volatility and the large quantities involved. With regard to solid wastes, those expected from fusion reactors are similar in quantity to those from today's fission reactors, but the wastes of high activity (e.g., the spent blanket in tokamak designs) contain only short-lived radioisotopes.

III. WASTE PROPERTIES

In the foregoing section, we have talked loosely of radwastes of low and high activity as two categories which should be treated differently as far as handling, storage and final disposal are concerned. However, it became clear that things are not so simple: some wastes of low activity are expected to contain traces of highly radiotoxic isotopes which set them apart from the rest. Thus, it is clear that the only sure way of categorizing radwastes is to determine exactly what isotopes are present in which amounts before deciding on the most appropriate disposal method. Ideally, this would have to be carried out systematically for all wastes from every source, but this is obviously a task which would far exceed all technical and economical possibilities. Add to this the complications introduced by the change in isotopic composition with time due to the different decay rates of the original radioisotopes and their daughters, and the problem is clearly insuperable. Today, the tendency in relation to final disposal is towards a pragmatic classification, based on estimates of the level of the potential health hazard of critical isotopes. These estimates in turn are based on assessments of the long-term behaviour of these isotopes in the different types of geological environment considered for the final respositories. Whether a particular batch of waste falls into a particular category is not assumed *a priori*, but rather judged separately in each case on the basis of the determined isotopic composition of the batch and the specific character of the repository site. For the purpose of discussing long-term disposal concepts, however, it is convenient to use a simple twofold subdivision based on estimates of the long-term health hazard associated with typical radwastes from different sources. Before outlining this classification, we first look at the different types of ionizing radiation and their health effects and at the main radioisotopes found in the wastes.

A. Ionizing Radiation

Elements consist of one or more isotopes with the same chemical characteristics (same number of protons and electrons) but with different atomic weights (different numbers of neutrons in their nuclei). Radioisotopes are isotopes which are unstable and spontaneously change into another isotope of the same element or into an isotope of another element. In so doing, they emit various types of ionizing radiation, radiation which on impact with other atoms or molecules in its path imparts enough localized energy to remove or add electrons (i.e., to ionize them). Some ionizing radiation is electromagnetic (i.e., similar to X rays) and some consists of high-velocity subatomic particles of various masses and charges. The spontaneous ejection of subatomic particles causes the radioisotope to decay and the decay rate (number of atoms ejecting particles in a certain small time interval) is proportional to the number of atoms of the radioisotope present. The constant of proportionality is called the decay constant and the decay curve (number of atoms versus time, see Fig. 29) is exponential and often expressed in terms of half-life, i.e., the time required for half the remaining atoms to decay. The number of atoms changing per second is taken as a measure of the radioactivity or a given amount of a radioisotope, conventionally expressed in terms of *curies* (1 Ci is the decay rate of 1 g of the radioisotope ^{226}Ra and equals 3.7×10^{10} atoms/sec, or becquerels). Also, the radioactivity of a given amount of radwaste is often given as an estimate of the number of curies (i.e., equivalent grams of ^{226}Ra), even though then many different radioisotopes are involved. However, because of the different types of ionizing radiation occurring in such a mixture, other parameters must be used to compare the potential hazards of different waste types.

Three main types of ionizing radiation can be distinguished: α, β, γ. α-Emitters radiate particles consisting of two protons and two neutrons (α-particles = helium nuclei), β-emitters radiate electrons (β-particles, with either + ve or − ve electrical charges), and γ-emitters radiate short-wavelength, high-frequency electromagnetic waves, usually together with one of the other radiation types. Neutron emission is also an important type of radiation, particularly in the cores of reactors, but it is not directly ionizing.

The different types of radiation have different capabilities of penetration and different effects on living tissues. In air, α-particles only penetrate a few millimetres, whereas β-particles penetrate a few metres and γ rays up to a few hundred metres. On the other hand, α-radiation is about 10 times as injurious to living cells as the same amount (energy equivalent) of β- or γ-radiation. This means that, for an equivalent amount of radiation, the occurrence most likely to be detrimental to health is the ingestion or inhalation of α-emitting radioisotopes (see Table I). The most common unit of dosage used to compare the biological effects of different types of radiation is the rem (roentgen equivalent man). A rem is the amount of any radiation which produces a biological effect equivalent to that resulting from an amount of γ-radiation equal to 100 ergs of energy transferred to 1 g of tissue ($= 1$ rad). The natural background radiation (cosmic radiation and radiation from naturally occurring radioisotopes) of a person living at sea level is of the order of 100 millirems/year, but it can reach double that amount at high altitudes or in areas where the Earth's crust contains high contents of radioisotopes. The additional radiation dosage due to man-made sources such as fallout, medicine, and nuclear power production) is variously estimated at 40–100 millirems/year in highly industrialized nations (Fig. 4). The aim of present radwaste disposal strategies in the United States is that the background radiation

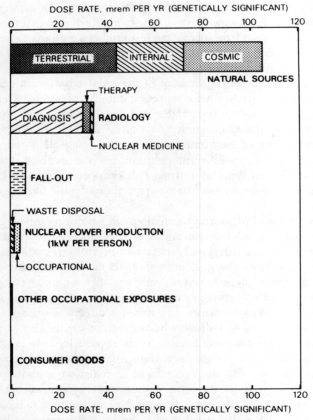

Fig. 4.   Annual genetically significant dose rate, as averaged through the whole population, based on statistics of the Nuclear Energy Agency of the OECD. (From Aikin *et al.*, 1977.)

should not be increased by, or the general population should never be exposed to, more than 5 millirem/year from that source.

B. Main Radioisotopes in Radwastes

Of all the known natural and man-made radioisotopes only a limited number are of importance for the problem of radwaste disposal. Most of those used in medicine, research, and industry are β/γ-emitters with very short half-lives, and the health hazard they pose becomes negligible after short periods of storage. The main concern derives from the long-lived radioisotopes produced in both military and commercial nuclear reactors (Table I). These can be subdivided into four main groups with different origins and characteristics: transuranics, uranics, fission products, and the group of light isotopes formed by neutron bombardment of the materials immediately surrounding the fuel rods.

The transuranics (atomic weights > 237) all originate by neutron capture and are all radioisotopes which do not normally occur in nature. The long-lived isotopes of concern are mainly α/γ-emitters and therefore very radiotoxic if inhaled or ingested. Their main representative is ^{239}Pu, which makes up 0.5% by weight of spent fuel. The transuranics slowly decay (over thousands of years) to a whole spectrum of uranics (atomic weights between 207 and 238) and join those members of this group formed by the decay of the original uranium. These radioisotopes and other daughters of the uranium decay series are found in nature (decay products of natural uranium) and the isotopes with long half-lives are again all α-emitters. Uranium isotopes themselves have very low specific activities (corresponding to their very long half-lives) and thus their radiotoxicity is very low. The most important uranic from the point of view of health hazard is ^{226}Ra and its highly radioactive, short-lived daughter, ^{222}Rn. Radon is a gas and is therefore the main danger to health in uranium mines and mills and in the neighbourhood of uranium mill tailings dumps. The fission products in spent reactor fuel (atomic weights between 85 and 155: "split" uranium atoms) are all β/γ-emitters and most are relatively short-lived (high specific activities). They are present in such quantities in spent fuel (0.5%) that at first they completely dominate the energy release, generating large amounts of heat and requiring heavy shielding. The fission products causing most concern are ^{90}Sr and ^{137}Cs, with half-lives of about 30 years, and ^{99}Tc, together with the volatile radioisotopes ^{85}Kr and ^{129}I. Finally, the fourth group of light radioisotopes (atomic weights < 60) do not originate in the fuel itself, but by neutron bombardment of stable isotopes in the cladding hulls and coolants surrounding the fuel rods. These are β-emitters, the most important being ^{60}Co and ^{55}Fe in the hulls and ^{3}H, which also occurs as a fission product in the spent fuel, in the coolant water.

Table 1. Main Radioisotopes with Half-Lives Longer than 1 Year Occurring in Radwaste from the Nuclear Fuel Chain[a]

Element	Radioisotope	Half-life (years)	Type of rad.[b]	Origin[c]	% Weight in spent fuel	Specific activity (curies/g)
Tritium	³H	12.3	b	FP and NC	—	3600
Carbon	¹⁴C	5730	b	NC	—	—
Iron	⁵⁵Fe	2.7	EC	NC	—	—
Cobalt	⁶⁰Co	5.3	b+y	NC	—	1140
Nickel	⁶³Ni	100	b	NC	—	—
Fission products						
Krypton	⁸⁵Kr	10.8	b+y	FI	0.08	39
Strontium	⁹⁰Sr	28	b	FI	0.09	140
Zirconium	⁹³Zr	1,000,000	b	FI	—	0.0026
Technetium	⁹⁹Tc	210,000	b	FI	—	0.017
Ruthenium	¹⁰⁶Ru	1	b	FI	—	2500
Palladium	¹⁰⁷Pd	7,000,000	b	FI	—	—
Antimony	¹²⁵Sb	2.7	b+y	FI	—	—
Iodine	¹²⁹I	16,000,000	b+y	FI	0.3	0.00016
Cesium	¹³⁴Cs	2.1	b+y	FI and NC	—	1350
Cesium	¹³⁵Cs	2,000,000	b	FI	0.28	0.00077
Cesium	¹³⁷Cs	30.2	b	FI	—	91
Promethium	¹⁴⁷Pm	2.6	b	FI	—	910
Samarium	¹⁵¹Sm	90	b+y	FI	—	28
Europium	¹⁵⁴Eu	16	b+y	FI and NC	—	141
Europium	¹⁵⁵Eu	1.8	b+y	FI	—	1280

Uranics		Half-life (years)	[b]	Source[c]		
Lead	210 Pb	22.3	b	ND(^{222}Rn)	—	—
Radon	222 Rn	.01	a	ND(^{226}Ra)	—	—
Radium	226 Ra	1622	a+y	ND(^{230}Th)	—	1
Actinides						
Thorium	229 Th	7340	a+y	ND(^{233}U)	—	—
Thorium	230 Th	75,400	a	ND(^{234}U)	—	—
Uranium	234 U	247,000	a+y	ND(^{238}U)	—	0.0062
Uranium	235 U	710,000,000	a+y	NS	0.76	0.000002
Uranium	236 U	24,000,000	a+y		0.46	0.00005
Uranium	238 U	4,510,000,000	a+y	NS	94.2	0.0000003
Transuranics						
Neptunium	237 Np	2,100,000	a+y	NC & TD(^{241}Am)	0.05	0.0007
Plutonium	238 Pu	87	a+y	NC & TD(^{242}Cm)	0.02	17.5
Plutonium	239 Pu	24,110	a+y	NC	0.53	0.062
Plutonium	240 Pu	6580	a+y	NC	0.22	0.23
Plutonium	241 Pu	14.4	a+y	NC	0.10	114
Plutonium	242 Pu	380,000	a+y	NC	0.04	0.0038
Americium	241 Am	445	a+y	NC and TD(^{241}Pu)	—	32.2
Americium	243 Am	7650	a+y	NC	0.02	0.18
Curium	244 Cm	18	a+y	NC	—	83.3
Curium	245 Cm	9300	a+y	NC	—	—

[a]Main sources: Organization for Economic Cooperation and Development, 1977; Lipschutz, 1980; Herrmann, 1983.

[b]a, α; b, β; y, γ.

[c]EC, orbital electron capture; NC, neutron capture; Fl, fission; ND, natural decay (daughter of . . .); NS, natural source; TD, transuranic decay (daughter of . . .).

The removal of tritium from coolant and process fluids so that these can be safely released to the environment is one of the most problematic aspects of waste treatment, both in nuclear reactor parks and reprocessing plants.

C. Waste Classification

The original classification of radioactive waste used (low-active and high-active) was based on the total amount of radioactivity in a given volume of waste and was geared to the needs of developing handling and storage techniques which limited the exposure of individual workmen and of the general population to ionizing radiation. Since maximum permissible dose levels continue to be controversial and since they vary from country to country, the interface values between low-active and high-active waste are not generally agreed upon. A whole range of limiting values are in use, varying between $\frac{1}{10}$ and 10,000 Ci/m^3, with most lying in the range of 100 to 1000 Ci/m^3. Because of these uncertainties, many countries distinguish a category of intermediate-active waste, covering the central part of the range (Fig. 5). Low-active waste can be handled directly without the occupational dosage exceeding specified values (e.g., in the United States, low-active = < 10 Ci/m^3, whole body occupational exposure limit = 5 rem/year). High-active waste must be shielded and/or handled by remote control if the maximum permissible dose is not to be exceeded.

The problem with this classification is that it does not provide a guideline for waste disposal, only for handling and interim storage. It considers the health of present generations but not of those far into the future. Since long-term disposal rather than short-term storage is the ultimate aim of radwaste management, the tendency now is to subdivide wastes into low-level and high-level, based on estimates of the potential health hazard at any time after disposal. To a first approximation, this depends on the amount of long-lived α-emitters; release to the biosphere from a carefully sited and designed repository should only occur after very long periods of time, when the only radioisotopes remaining will be uranics and transuranics (Table I). Once again, however, it is difficult to reach a consensus on future permissible exposure levels, and it is practically impossible to define a general low-level/high-level interface without reference to the specific repository in question. One possibility is to take the α-particle emission rate of uranium ore as the interface value ($10^{-2}-10^{-3}$ Ci/m^3) (see Fig. 5). This is the present procedure in the United States, whereby the wastes with only marginally higher levels of α-radiation are put in a special intermediate category, called variously TRU wastes, α-wastes, or α-contaminated wastes. Another possibility is to use model calculations of radioisotope release from a general type of repository now in use to define low-level waste concentration limits such that conserv-

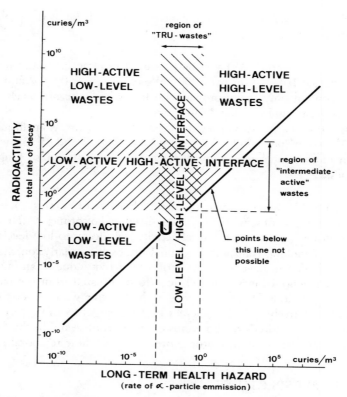

Fig. 5. Categorization of radwaste on the basis of total radioactivity (low-active/high-active) and potential health hazard (low-level/high-level). Health hazard is assumed to be related to the rate of α-particle emission in the wastes (potential ingestion or inhalation hazard), which in turn is related to the content of transuranic and uranic radioisotopes. There is no agreement on the exact positions of the interfaces. The range of values covering the low-active/high-active interface indicates the region of so-called intermediate-active wastes. The range of values covering the low-level/high-level interface indicates the region of so-called "TRU wastes" (United States) or "α-contaminated" wastes (IAEA). Not included is a line representing the interface between radioactive and nonradioactive (innocuous) wastes, since this has not yet been defined precisely. The position of uranium ore is indicated by the letter U. (Compiled from Bürgisser *et al.*, 1979; Heintz, 1980; Carter *et al.*, 1979; Cohen and King, 1978 and International Atomic Energy Agency, 1981.)

ative guidelines for radiation exposure to the general public could never be exceeded. A recent study of this type indicated a low-level/high-level interface value of 1 Ci/m³ for α-radiation (Fig. 5).

From this short discussion it is clear that the question of waste categorization involves a whole spectrum of uncertainties which are unlikely to be resolved in the near future. For the purposes of this book, therefore, I will assume that there will be two basic waste categories as far as the problem of long-term disposal is concerned:

1. Low-level waste (LLW) will be produced in large quantities [order of magnitude 10^3–10^5 m^3 per 1000-MW reactor/year (Figs. 2 and 3)] and will contain only radioisotopes with low specific hazard levels and/or short decay times in appreciable amounts.
2. High-level waste (HLW) will be produced in relatively small quantities [10^0–10^2 m^3 per 1000-MW reactor/year (Figs. 2 and 3)] but will contain radioisotopes with very high hazard potential and/or long decay times.

LLW can be highly radioactive initially (high-active LLW) (Fig. 5), but only because of a high content of very short-lived radioisotopes. After a short period of storage, however, the total activity will decay to low levels which allow direct handling, transport, and final disposal. Similarly, on the above definition, HLW can have a low level of total activity, due to an initial lack of short-lived radioisotopes, but still need to be isolated from the biosphere for very long periods of time. In general, high-active waste reverts to low-active waste relatively quickly, whereas HLW changes to LLW very slowly, often only after hundreds of thousands or millions of years. In the final section of this chapter we look a little more closely at these temporal changes.

IV. TIME PERSPECTIVES

Throughout the preceding discussion, time has been a factor implicitly involved at several points, all arising from the basic fact that radioisotopes decay with time (see half-lives, Table I). In this final section, we treat some aspects of the time factor more explicitly, to emphasize its importance for the geological aspects of radwaste disposal. The main question is: for what time periods must radwastes be confined or isolated before they become innocuous and can be released to the biosphere with no greater hazard to health than common naturally occurring radioactive materials?

Radwaste contains mixtures of various radioisotopes, each decaying at a specific rate given by its half-life. The total activity of radwaste and the dominant radiation type also change with time in a complicated way which can only be predicted if the exact radioisotope composition is known. A rule of thumb of the early nuclear industry was that the confinement or isolation time of radwastes should be at least 10 times the half-life of the longest living of the dominant isotopes. After 10 half-lives, the radioactivity of any isotope decays to less than a thousandth of its original value. For low-active waste, this rule gave at least 300 years (dominant isotopes ^{137}Cs and ^{90}Sr) (see Table I) and for high-active waste or spent fuel, at least 250,000 years (dominant isotope ^{239}Pu) (see Table I). This rough method is obviously unsatisfactory, but it gives values of the same order of magnitude as more sophisticated analyses.

If the precise radioisotope composition is known, it is possible to calculate the variation of various parameters with time, of which the most important for radwaste disposal strategies are heat generation, total radioactivity and hazard index. Taking spent fuel as the most hazardous type of radwaste which may need to be isolated, its heat output drops from hundreds of kilowatts per ton on discharge from the reactor to just over 1 kW/ton after 10 years (Fig. 6). During this period of very high heat generation it must be stored at the reactor site in special pools with constant cooling and surveillance. At the end of the same period, however, its total radioactivity is still high, more than 400,000 Ci/ton, due to the β/γ-radiation of the fission products. After about 300 years of radioactive decay, when the radioactivity has sunk to about 3500 Ci/ton, a significant change takes place. As the relatively short-lived fission products disappear, the radioactivity becomes mainly due to the α-radiation emitted by the much longer-lived uranics and transuranics. This drops much more slowly over the succeeding millenia, until it finally reaches the same level as the amount of ore required to produce the fuel in the first place,

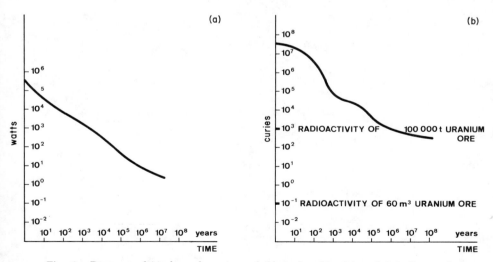

Fig. 6. Decrease of (a) thermal output and (b) total radioactivity of 60 m^3 spent fuel (compared with the amount of uranium ore used to produce the fuel and the same volume of uranium ore) with time. (Compiled from Lipschutz, 1980; Hermann, 1983 and National Academy of Sciences, 1983.)

several hundred thousand years from now. However, the spent fuel would then represent a concentrated source of radioactivity as compared with the diffuse distribution of uranium throughout thousands of tons of ore. If the spent fuel is compared after 1 million years (m.y.) with the same volume of ore, it is at least 10,000 times more radioactive (Fig. 6).

The parameters of heat generation and total radioactivity are not good indicators of the danger posed by radwaste. This can be better assessed by a theoretical hazard index, defined as the volume of water which would have to be used to dilute a given quantity of waste so that the water could be used as drinking water (Fig. 7). The first part of such a hazard index/time curve mimics the shape of the radioactivity curve. After 1000 years, however, most estimates indicate a long period of leveling off. This is the period

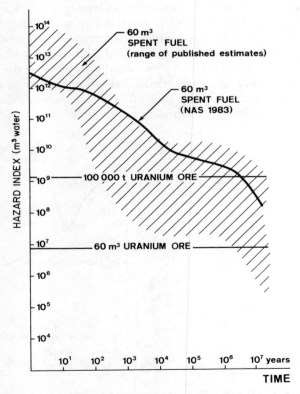

Fig. 7.   Decrease in hazard index (water dilution volume) of spent fuel with time, compared with the hazard index of the amount of uranium ore required to produce the fuel and the hazard index of the same volume of uranium ore. (Compiled from Jet Propulsion Laboratory, 1978 Lindblom and Gnirk, 1982 and National Academy of Sciences, 1983.)

when the hazard derives from the content of α-emitting transuranics, which remains relatively constant throughout this time. It is also the period when the spent fuel has the same hazard potential as the total amount of ore used to produce it. On a volume-to-volume basis, however, the spent fuel remains 100 times more hazardous than uranium ore throughout this period.

From this brief discussion of the time factor, some general guidelines can be deduced which will be of use as orientating numbers. For LLW, the time frame to be used for confinement will be between 100 and 1000 years, not only from the point of view of the period for which social commitment can be made, but also from the point of view of radioactivity and hazard index (disappearance of unstable fission products). For HLW, isolation from the biosphere for more than 1 m.y. may have to be considered in the case of spent fuel, and for HLW from reprocessing, isolation times of 10^5–10^6 years are generally used as guidelines for developing appropriate disposal concepts.

V. CONCLUDING REMARKS

This brief description of radioactive wastes was intended to give some idea of the wide spectrum of materials involved and the range of different quantities and sources. The characterization and management of these wastes is a complex technical problem which will not be treated further in this book. From the point of view of long-term disposal, which is the aspect of waste management in which the earth sciences play a dominant role, it is sufficient to assume that there will be two main categories of radwaste, low-level waste (LLW) and high-level waste (HLW), depending on their content of long-lived α-emitting radioisotopes. Because of the large quantities, LLW will require management procedures aimed at confinement, i.e., active containment with some degree of social commitment (surveillance, monitoring, keeping of site records, security), possibly involving controlled or predictably low release of radioisotopes. After a period of a few hundred years (the time period over which a social commitment can probably be upheld and during which the radioactivity will have dropped to negligible levels) it should be possible to release LLW to the environment with the assurance that radiation exposure guidelines for individuals or populations will not be exceeded under any circumstances. For HLW, the radioisotope content is such that it could not be released to the environment for a very long period of time (far longer than that for which social commitment can reasonably be guaranteed) without a high risk of exceeding conservative radiation exposure guidelines. HLW requires isolation, i.e., management procedures aimed at complete containment of the waste within the repository, with no expected release to the biosphere and no need for a commitment of society except for an initial period. For the purposes of this book, the category HLW includes high-level reprocessing wastes, spent fuel, and some proportion of the low-active reprocessing wastes in which the content of transuranics exceeds a specified value (high-level TRU wastes). The most important differences between these types of HLW will be pointed out at the appropriate places.

SELECTED LITERATURE

General sources and reviews: U.S. Energy Research, Development, Administration, (ERDA), 1976; Romer, 1976; Aiken *et al.*, 1977; English *et al.*, 1977; Organization for Economic Cooperation and Development (OECD), 1977; Lipschutz, 1980; International Atomic Energy Agency (IAEA), 1980a; U.S. Department of Energy (DOE), 1982b; Burton *et al.*, 1982; Herrmann, 1983; National Academy of Sciences (NAS), 1983 (pp. 24–42).

The nuclear fuel chain ("once-through" option): OECD, 1977 (pp. 89–106); DOE, 1979b; Carter *et al.*, 1979 (pp. 55–78); Phillips *et al.*, 1979; Stockholm International Peace Research Institute (SIPRI), 1979 (pp. 91–103); Deffeyes and MacGregor, 1980; Swedish Nuclear Fuel Supply Co., 1983 (pp. 3:1–3:17); Kesson and Ringwood, 1983.

Nuclear fuel cycles (reprocessing options): Union of Concerned Scientists (UCS), 1975 (pp. 219–275); American Physical Society, 1977; Kreiter *et al.*, 1977; OECD, 1977 (pp. 107–126); Tera Corp., 1978; Carter *et al.*, 1979 (pp. 89–102); SIPRI, 1979 (pp. 105–111); Heintz, 1980; IAEA, 1983g.

Future radwaste sources: Kaser *et al.*, 1976; Gilmore, 1977 (pp. 25–34); OECD, 1977 (pp. 127–136); Lipschutz, 1980 (pp. 50–54); IAEA, 1981 (pp. 153–163), 1983a; Vogler *et al.*, 1983; Dougherty and Adams, 1983.

Ionizing radiation and health risks: Aikin *et al.*, 1977 (pp. 25–30); Cohen, 1977; Sax, 1979 (pp. 139–188); Upton, 1982.

Main radioisotopes in radwaste: Gilmore, 1977; OECD, 1977; Lipschutz, 1980; Herrmann, 1980.

Waste classification and isolation times: Gera, 1974; English *et al.*, 1977; Cohen and King, 1978; Bürgisser *et al.*, 1979 (pp. 18–23); Carter *et al.*, 1979 (pp. 103–125); Smith *et al.*, 1980; de Marsily and Merriam, 1982 (pp. 9–24); Herrmann, 1983 (pp. 42–47).

CHAPTER 2

Radwaste Disposal

I. INTRODUCTION

In the previous chapter, we discussed what radwaste is, how it is produced, and what it contains, in order to see what requirements are to be placed on any disposal system. Here, we shall look at the disposal systems themselves, some of which have been in use, successfully or unsuccessfully, for many years, and some of which are still in the conceptual or developmental stages. All systems are complex strategies with numerous components based on some combination of basic philosophies, depending on basic choices concerning the long-term aims. These can be briefly summarized as the answer to two basic questions: containment or dispersal? and confinement or isolation?

Containment implies keeping the wastes within prescribed boundaries, generally within the boundaries of the repository (the waste packages and the immediately surrounding natural medium), whereas dispersal involves releasing the wastes to the natural environment in a form which ensures rapid spreading and dilution to innocuous concentration levels. The large quantities of radwaste being produced, the existence of physicochemical and biological concentration mechanisms and the ever-growing environmental pressures from other waste materials, such as heavy metals and phosphates, however, have combined to make dispersal unacceptable as a primary aim. At the present time, most disposal strategies aim at containment but many include a dispersal philosophy as a last resort (i.e., for the event of containment failure, an environment with a favourable dispersal function may be chosen).

Confinement implies active containment of waste within the repository and active prevention of releases to the biosphere which may cause significant health hazard. It may also involve controlled or predictably low dispersal outside the repository, subject to constant surveillance and monitoring and subject to the possibility of intervention if such dispersal should lead to excessive concentrations. Confinement implies a social commitment (perpetual care) and the possibility of retrieval of the waste, if necessary or opportune. In contrast, isolation implies no social commitment, no possibility of retrieval, and no chance of intervention if something goes wrong. Isolation involves a passive system of containment (usually in the form of a whole series of barriers, each of which would have to fail before a release could take place) and a reliance on the geological system around the repository to prevent dispersal or to allow dispersal at such low rates that a release to the biosphere can be discounted. Disposal strategies for low-level waste (LLW) generally emphasize confinement whereas those for high-level waste (HLW) are always aimed at achieving isolation (Fig. 8).

All the disposal concepts discussed below are based on a knowledge of the type of radwaste, the form it is in after treatment, and the mix of basic

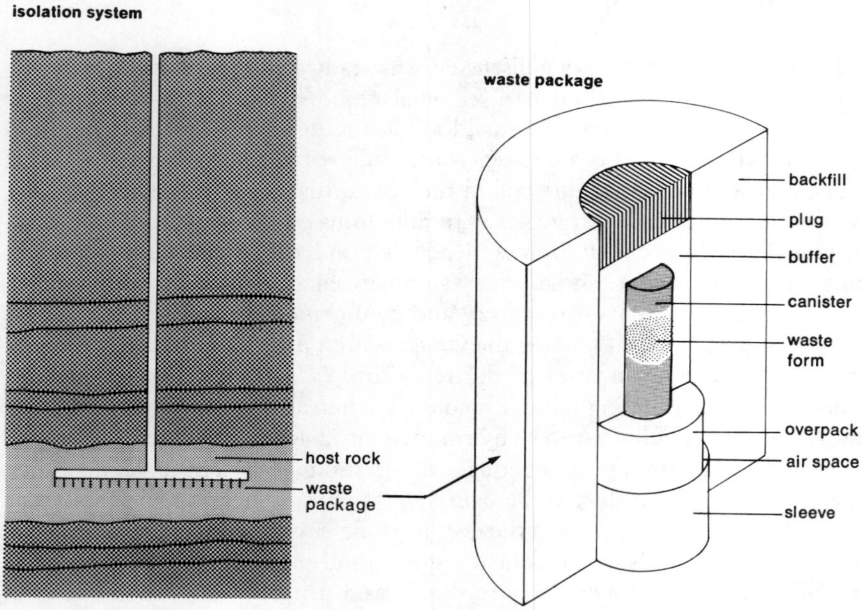

Fig. 8. The isolation system, illustrated with reference to the concept of deep underground disposal, and the different components of the waste package (artificial barriers to radionuclide migration). (From Klingsberg and Duguid, 1982, by permission of Sigma Xi, The Scientific Research Society.)

philosophies (confinement–isolation and containment–dispersal). First, we shall look at LLW, in both liquid and solid form. Such wastes have been accumulating in large quantities since World War II as a by-product of the weapons industry, both from uranium mining and from industrial processes, and have recently been joined by wastes from the development of nuclear energy. LLW disposal programmes have been running for many years and have a wide base of practical experience, not always positive, so they are far past the conceptual stage. Even so, many difficulties remain and the large quantities involved make the problem particularly pressing. In the final section of this chapter, disposal concepts for HLW will be outlined. Here, we will be dealing with suggestions for systems which are judged to hold out some hope of success at the present time. None of these systems have yet been tested, although many are the focus of intensive current research. The world's HLW is at present in interim storage, mainly in the form of untreated spent fuel, awaiting the solution of the many technological, administrative, and political problems. However, in comparison to LLW, the quantities of HLW are relatively small and interim storage would be physically possible for many decades, until acceptable final disposal programmes have been worked out. The emphasis in this chapter is on a broad overview of those concepts and programmes which need to be assessed from a geological viewpoint: a more detailed evaluation of the earth science aspects of these concepts and practices is the subject of Part II.

II. DISPOSAL OF LIQUID LLW

Because containment is the principal aim of most present waste disposal strategies, it is usually proposed to convert all wastes to a solid form as one of the first steps. Up to the mid-1970s, however, large volumes of liquid LLW from nuclear weapons manufacture were "disposed of" in the United States by rather haphazard procedures involving direct release to rivers and to the groundwater. At the present time, similar activities are still normal practice in some countries, e.g., release of low-level liquid effluents directly to the sea by Britain and France or the injection of liquid HLW into deep rock formations by the Soviet Union. Such strategies are either true dispersal systems or, because of the mobility of fluids in the Earth's crust, risk becoming dispersal systems at some future time. Some systems can, however, be successfully used locally for containment.

A. Ground Percolation

At the Hanford Reservation (Washington, the United States), the coolant water from the original cluster of eight nuclear reactors was released directly into the Columbia River until the reactors were shut down in the mid-1960s.

Since then, the liquid LLW from new reactors and the reprocessing plants on the reservation has been directed to underground cribs or open ponds from which it percolates down through soils, sands, and gravels to the groundwater surface about 100 m below (Fig. 9). During percolation, the contained radionuclides are selectively retarded by contact with the materials through which the waste seeps (see Chapter 5, Section III), a system similar to that used to purify river water by allowing it to percolate through soil, which is a routine procedure used by many water authorities. Instruments in wells around the cribs monitor the radionuclide contents at various levels and in the groundwater itself, and the crib is deactivated as soon as concentrations of any of the radionuclides exceed a small fraction of the concentration guides for drinking water. Since the groundwater flow is towards the Columbia River, some radionuclides have been added to the river also from this source, and the concentration of radionuclides in the river water and bottom sediments has been raised considerably over the years. Although these concentrations are still far below recognized limits and although all releases are strictly controlled, it is unlikely that this disposal strategy (essentially controlled dispersal) will ever become generally acceptable.

B. Deep Well Injection

Several types of deep well injection procedures, similar to those used for disposing of large volumes of dilute liquid industrial wastes, have been proposed, and in some cases tested, for liquid LLW disposal. The principle is to transfer the liquid via boreholes to a situation deep in the Earth's crust where the waste would be contained in the same way that petroleum, water, and gas have been contained in natural oil traps for millions of years. The liquid wastes would be injected into a porous host rock surrounded by suitable confining beds, either displacing the water or air already present or creating their own space by being injected under very high pressures (hydraulic fracturing). In a pilot project of this type in the Soviet Union, 1.2×10^6 m^3 of liquid LLW (total radioactivity 50×10^6 Ci) were disposed of in sandstones at a depth of 350 m between 1960 and 1973 and the effects observed in an array of instrumented control wells. As in the case of ground percolation at Hanford, radionuclide migration was retarded by ion exchange with the host rock (see Chapter 5, Section III,E), but some other effects were rather alarming (here, that rock temperatures were raised from 12 to 54° C over 3 years). Also, worldwide experience with deep injection of industrial wastes has shown that migration paths and degrees of confinement are difficult to predict accurately (poorly known subsurface geological conditions) and that the injection process often has undesirable side effects (land surface movement, small

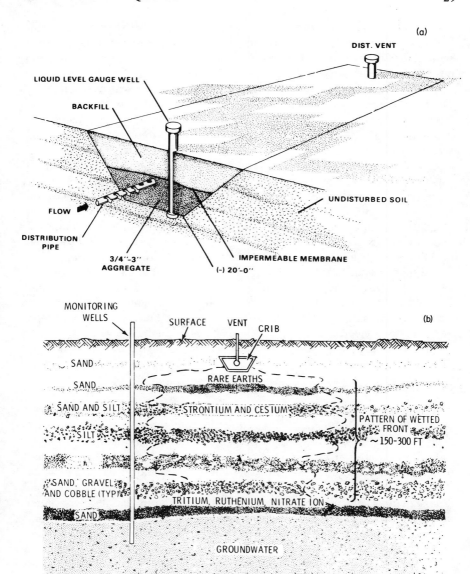

Fig. 9. Ground percolation of liquid reactor and reprocessing wastes at Hanford Reservation, Washington. (a) Typical disposal crib; (b) movement and fractionation of radionuclides in the unconsolidated sediments below such a crib. (From National Academy of Sciences, 1978.)

earthquakes). Since liquid LLW must be disposed of at its point of origin, and since nuclear reactor and reprocessing plant sites are not usually chosen with favourability for deep well injection in mind, it is unlikely that this process will be used extensively.

One example of a liquid radwaste injection scheme which has received official approval is that carried out at an experimental farm on the Nevada Test Site (Nevada, the United States). There, liquid LLW, contaminated with ^{238}Pu, is being disposed of by pumping it down a 800-m-deep borehole, which was originally bored as an observation well after one of the underground nuclear explosions carried out in the area. The amounts involved are relatively small (about 25,000 gallons up to 1977) and the addition to the enormous underground inventory of radionuclides on the site is insignificant. Also, the hydrogeology of the area is well known and possible releases to the environment have been continuously monitored since bomb testing ceased and will be for the foreseeable future. Up to 1977, the waste disposal activity had had no effect on the measurements. Another example is the application of deep well disposal techniques to the problem of disposing of large volumes of liquid LLW generated by the *in situ* uranium mining industry in South Texas. Apart from the United States in these limited areas, the only country following this strategy as the main thrust of its radwaste research, both low-level and high-level, is the Soviet Union.

C. Grout Injection

A variation on the theme of disposal of liquid wastes is the injection of grouts, which subsequently solidify, into the underground. This system has been in general operation at the Oak Ridge National Laboratory (Tennessee, the United States) since 1966, after 7 years of nonradioactive testing and under continuous monitoring from a system of observation wells. Up to 1978, about 11,000 cubic metres of grout had been disposed of in this way (total radioactivity at injection 640,000 Ci) and the programme has been continued up to the present time. The grout, a mixture of concentrated LLW liquids (with activities up to 6000 Ci/m^3) and cement solids, is injected under high pressure at depths of 240–290 m, where it creates its own space in the form of narrow, subhorizontal fractures in an extremely impervious rock, shale (see Chapter 7, Section IV,A). There it solidifies as sheetlike bodies at various levels (Fig. 10) and the radionuclides are effectively immobilized in an environment virtually free of circulating underground water which could cause leaching. A similar system is proposed in West Germany, where the siting of the projected fuel enrichment and reprocessing plant is being based on the site's favourability for the pumping of radioactive grouts into specially prepared underground cavities in salt domes.

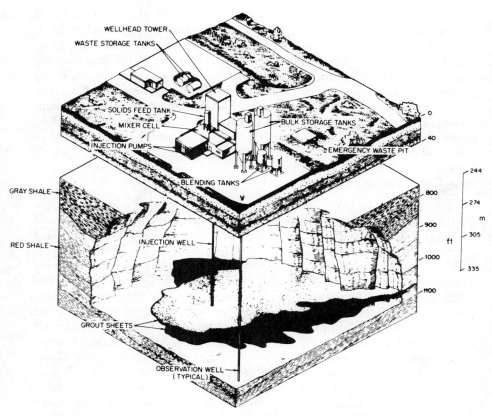

Fig. 10. Perspective view of the grout injection facility at the Oak Ridge National Laboratory, Oak Ridge, Tennessee. (From International Atomic Energy Agency, 1983b.)

III. DISPOSAL OF SOLID LLW

Although under certain unusual circumstances the direct disposal of liquid LLW may be possible, liquids are in general too mobile to guarantee confinement over long periods of time. Also, liquid wastes are only part of the LLW inventory from any nuclear plant; in addition, there is a wide range of contaminated solid items (including equipment, components, filter residues, and clothing). Hence, the most common procedure for LLW is to convert all liquid wastes to solids by evaporation and to strive towards an integrated strategy for solid LLW disposal. All such strategies include a combination of treatment and packaging procedures, many of them similar to those used for hazardous, nonradioactive wastes, aimed at reducing the volume and particle size, increasing the long-term stability, and producing manageable, standard-sized units. The typical end product is a canister made of steel, concrete, or some other inert, corrosion-resistant material, containing particulate LLW embedded in a solid matrix of concrete, bitumen, glass, or some other monolithic solid of low dispersability. Such LLW canisters are routinely disposed of around the world in three different geological environments: on the ocean floor, in shallow trenches or pits and in underground rock caverns. These three disposal methods are outlined briefly below. The problem of uranium mill tailings, which cannot be packaged because of the huge volumes involved and which have been managed rather haphazardly in the past, is treated at the end of this section.

A. Controlled Sea Dumping

Dumping solid LLW in the sea is, as the name implies, a haphazard procedure, which, at least during the early years, can hardly be called a disposal strategy (philosophical basis—"out of sight, out of mind"). Uncontrolled dumping of this type was carried out by the United States, Britain, and possibly other countries regularly from 1946 up to 1969. Controlled dumping under the auspices of the Organization for Economic Cooperation and Development (OECD) Nuclear Energy Agency started in 1967 and continues until 1983. Under this international programme, a total of about 100,000 tons of packaged solid LLW have been sunk in 4000–5000 m of water in a designated disposal area in the northeastern Atlantic. The programme is being accompanied by research and monitoring aimed at defining the actual effects of the dumping, as opposed to the model predictions used earlier. Similar activities have been proposed by Japan for a site in the western Pacific. Since the average life of most canisters under high pressure in sea water is estimated variously at 20–100 years, and since the resistance of the different matrices to dissolution under such conditions is unknown, sea dumping is essentially a dispersal rather than a containment strategy. As such, it has been discontinued by many countries (including the United States, West Germany, and Sweden) and it has been condemned by the European Parliament (leading to the recommendation of a 2-year moratorium by the London Conference of February 1982). A temporary end to dumping activities was achieved in 1983 by industrial action in the United Kingdom. Sea dumping will probably be replaced by land-based, national confinement procedures as soon as these have been developed.

B. Shallow Land Burial

Disposal of solid low-active waste in shallow trenches backfilled with soil and rubble has been practised extensively by the United States since the initiation of atomic weapon manufacture in the 1940s. The way in which this was carried out in the early years is generally recognized as a notorious example of poor waste management. Little account was taken of the types and amounts of waste, the location of the sites, or the effects on the environment. Since the U.S. Atomic Energy Commission was disbanded in 1974–1975, control and monitoring of the old sites has been instituted by the U.S. Dept. of Energy to provide forewarning of dangerous radionuclide releases, and the commissioning of new sites has been strictly regulated. In fact, the tendency today is to define LLW as waste which in the solid form can be disposed of by this method at a specific site such that the guidelines for radiation exposure to the general public will never be exceeded (see Chapter 1, Section III,C). This implies exact knowledge of the radioisotope content of the wastes, of the geology and hydrology of the site, of the rates and mechanisms of radionuclide migration from the repository, and of the hazard indices of the released radioactive species.

Under the old regime, $> 1 \times 10^6$ m^3 of solid low-active waste, much of it containing concentrations of transuranic elements which today would be too high for categorization as LLW ("TRU wastes") (see Fig. 5), were buried at at least 12 sites in the United States (see Chapter 5, Section IV). If the production of electricity from commercial nuclear reactors increases as planned, it is estimated that 1×10^6 m^3 of solid LLW will be accumulating per year by the end of the century. In spite of the bad experiences of the past, active confinement of these wastes, either in shallow land burial sites or in underground caverns (see below), seems to be the only reasonable way of dealing with such large quantities, both in the United States and in most other countries.

Modern disposal of solid LLW by shallow land burial (Fig. 11) owes much to the failures of the past and to the development of sanitary landfill methods for urban waste disposal over the last 15 years. Improvements include better waste treatment (volume reduction, immobilization, packaging, segregation of waste types, and description of radionuclide inventory), better engineering of trenches (lining, roofing, drainage systems, and special fill materials), and better siting, with more stringent scientific and administrative controls (scientific site selection procedures, monitoring, contingency plans, regulations and guidelines, and record keeping). Also, the needs of safe LLW disposal will probably become a factor in the siting of future nuclear facilities themselves, since off-site transport of large quantities of LLW is not considered practical. Countries in which this disposal method was, is, or is projected to be, the main method of solid LLW disposal include the United States, the Soviet Union, Britain, France, Denmark, and Czechoslovakia.

C. Disposal in Mined Cavities

Although shallow burial in the soils and unconsolidated sands and clays which cover much of the land surface is the least expensive disposal strategy,

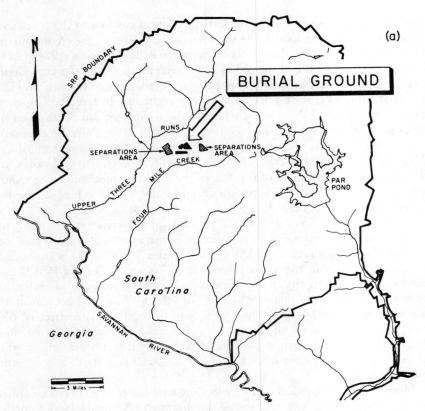

Fig. 11. Shallow land burial at the Savannah River Plant, South Carolina. (a) Map of the SRP reservation, with location of the burial ground; (b) plan of the trenches for different kinds of radwaste. This site belongs to the group of eastern United States burial grounds (humid climate) (see Table II) but is well drained and shows no radionuclide migration out of the trenches. HL, high β-γ waste; LL, low β-γ waste; TA, transuranium α waste; ST, storage; ▨, Spanish soil; ■, solid waste. (From United States Energy Research Development Administration, 1977.)

many countries are considering as an alternative deep burial in solid rock, in existing or specially constructed mined cavities. A typical design places the repository at depths of 100–200 m below the surface, in abandoned mines which are known to be dry or in rock masses which are thought to allow little water circulation (e.g., soft plastic rocks like salt or shale, or hard rocks like granite, which show a lack of cracks and fractures). If water exclusion from the repository could be guaranteed for the length of time required by the particular type of waste being deposited, a high degree of safety could be attained at such depths. However, because minimization of transport distance

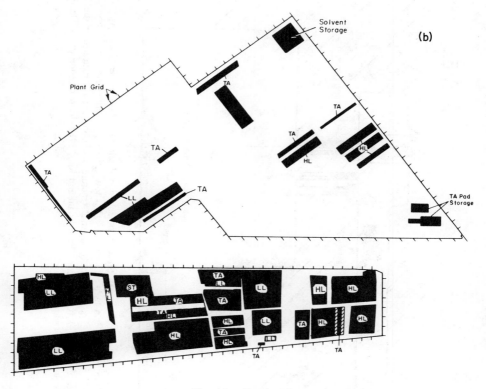

Fig. 11. Continued.

is an important factor by LLW disposal, many less favourable rock types and geological situations are being investigated at the present time, the decision being governed by the rock masses available beneath the nuclear plant in question (e.g., basalt lava flows beneath the Hanford Reservation). The aim here is to define the hydrogeological parameters so precisely that release mechanisms and rates, migration pathways, and, ultimately, hazard potentials can be predicted for different waste compositions. For many sites, there seems little doubt that confinement becomes easier to achieve and maintain at such depths and that the increased initial costs can be justified by this increased reliability.

The only repository of this type which has been operational is in an abandoned salt mine at Asse, near Hannover, West Germany (Fig. 12). Canisters of low-active to intermediate-active waste with a total volume of 25,000 m³ were routinely deposited there between 1967 and 1978. They fill about 14 of the original salt caverns in the mine and these have now been backfilled

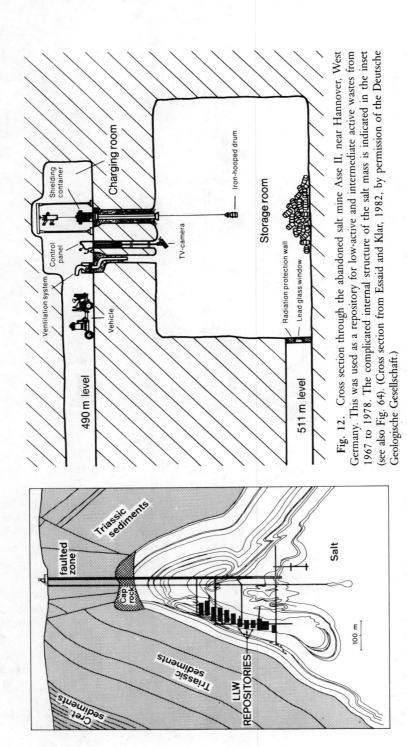

Fig. 12. Cross section through the abandoned salt mine Asse II, near Hannover, West Germany. This was used as a repository for low-active and intermediate active wastes from 1967 to 1978. The complicated internal structure of the salt mass is indicated in the inset (see also Fig. 64). (Cross section from Essaid and Klar, 1982, by permission of the Deutsche Geologische Gesellschaft.)

with salt and sealed, so the waste is essentially irretrievable (an isolation strategy). In 1978, the original licence ran out, and a new licence has not yet been granted under the stricter 1976 laws governing nuclear plants. Several demonstration and test facilities of this type are at present being established in the United States, including the Waste Isolation Pilot Plant (WIPP) in New Mexico (at 600 m depth in bedded salt) (see Fig. 50); the Basalt Waste Isolation Programme (BWIP) on the Hanford Reservation (at 50 m depth in basalt) (see Fig. 52); and the Sedan Crater project on the Nevada Test Site (at 100 m depth in a man-made crater formed by an underground nuclear explosion) (see Fig. 43). The possibilities of LLW disposal in abandoned mines are being investigated in several countries, including West Germany (Konrad iron mine), Czechoslovakia, Finland, and the United Kingdom, and conceptual designs for underground facilities have been developed by Sweden (ALMA project), Switzerland, West Germany, Belgium (Mol), and Spain (Sierra Albarrana).

D. Management of Uranium Mill Tailings

One of the most voluminous and problematic types of LLW is the residue left after processing uranium ores. Uranium extraction is carried out in large mills which are generally sited away from the mining operations, often servicing several mines. In 1982, the 30 operating mills in the western United States were processing about 35,000 tons of ore per day. Processing involves grinding the ore to a fine powder (milling), leaching the powder with acid or alkaline solutions to remove the uranium, filtering to remove the leached solids (tailings), precipitation of the uranium salts from the leachates, mainly as uranium oxide (yellow cake), and disposal or reuse of the barren liquids (tailings solutions). Processing methods vary from place to place, but in the end large volumes of LLW are produced (see Fig. 2), generally in the form of a slurry which is chemically aggressive and contains long-lived α-emitting radioisotopes such as ^{230}Th and ^{226}Ra (see Chapter 5, Section V). The slurries are led into large ponds where the water evaporates and/or percolates into the ground (Fig. 13). For every ton of yellow cake produced, between 250 and 1000 m^3 of solid LLW in the form of a fine sediment results. In the United States and Canada alone it is estimated that about 3×10^8 tons of uranium mill tailings exist at the present time and that this amount may double before the turn of the century. The radioactivity of the tailings is about the same as that of the original ore, and the main health hazard derives from the decay of ^{226}Ra to ^{222}Rn, which, although short-lived, is both gaseous and α-emitting (see Table I).

As with other types of radioactive waste, the early years of large-scale tailings production (1945–1965) were marked by poor management and hap-

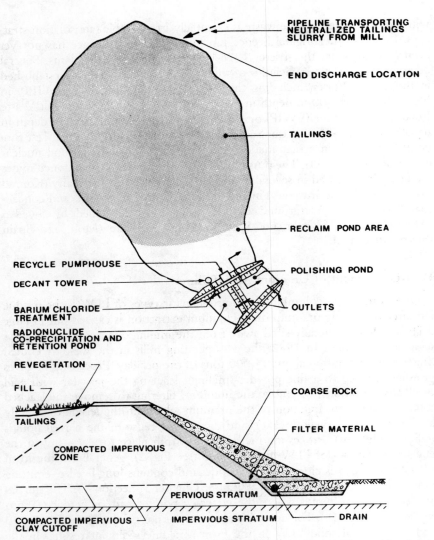

Fig. 13. Typical design of a surface tailings management system with provision for decommissioning. (From Hanney *et al.*, in Brawner, 1980, by permission of the Society of Mining Engineers of AIME.)

hazard disposal methods. This led to one of the few major environmental problems associated with the nuclear industry. Between 1952 and 1966, several hundred thousand tons of tailings were removed from abandoned piles by local citizens and builders for use as building material in the western United

States. In 1966, it was realized that the tailings being used for construction posed a potential radiological health problem to occupants and users of the structures involved, which included house foundations and walls, streets, and sidewalks, sewers, and public buildings. A comprehensive measurement programme showed by early 1970 that the radiation levels were above the then-accepted exposure guidelines and a remedial action programme was instituted, first at Grand Junction, Colorado (in 1972, to be completed by 1987, total cost $23 million) and then at many other localities (Uranium Mill Tailings Remedial Action Program, started in 1981, to be completed by 1990, estimated total cost $600 million). In the last decade, therefore, serious consideration has been given to the long-term management of these wastes, which, because of the very large volumes involved, cannot be packaged and are difficult to contain behind artificial barriers.

Present concepts assume that the wastes will immediately become part of the natural environment and aim at ensuring that this happens in such a way as to prevent bleed rates of contaminants which could exceed present regulatory levels (Fig. 38). Improvements are envisaged at several points in the process. Chemical separation steps may be inserted (concentrated sulphuric acid leaching, lime and/or barium chloride treatment), aimed at removing radium, thorium, manganese, heavy metals, and other contaminants before deposition in the tailings ponds or before release of effluents to the hydrosphere. Thermal stabilization of the tailings (sintering to 1100–1200°C) is being considered, since this strongly reduces the radon-emanating power, and chemical stabilization with other waste products (fly ash, kiln dust) has been suggested, in order to consolidate the wastes and make them resistant to wind erosion or water action. Many different containment barriers (impermeable bottom liners, improved drainage, engineered trench construction) and many different covering techniques are at present being tested. However, all improvements depend very heavily on local conditions of climate, ore quality, resource availability, population density, and economics.

IV. DISPOSAL OF SOLID HLW

The HLW of the world is at present in temporary storage, either as spent fuel at the nuclear power stations, or in a liquid or solid form at the reprocessing plants. Small amounts have been immobilized (converted to a chemically inhert solid form), mainly in test facilities and on a noncommercial scale, but none has yet been finally disposed of in the Western Hemisphere. This means that all disposal strategies for HLW are in the conceptual stage and that alternative strategies must be considered, depending on how the presently stored materials will be treated before disposal. Two main alternatives are usually considered (see Chapter 1):

1. The "once-through" option, under which the spent fuel from the power stations will be stored and cooled until its heat-generating capacity reaches a sufficiently low level and then encapsulated in a container of inert material or converted to an inert form before emplacement in the repository (Fig. 2)

2. The recycling option, under which the spent fuel will be reprocessed to extract the uranium and plutonium for reuse as reactor fuel, thereby producing highly radioactive liquid waste streams which will be stored and cooled until their heat-generating capacity reaches a sufficiently low level that they can be solidified, immobilized, and encapsulated for disposal (Fig. 3).

The once-through option would produce about 10–20 times as much solid HLW as the recycling option and yields wastes with an equally high initial radioactivity but a higher content of long-lived α-emitters. In both cases, disposal strategies aimed at guaranteeing isolation for hundreds of thousands of years must be followed. In either case, the amount of solid HLW is ~ 10–1000 times less than the amount of solid LLW generated at the same time.

HLW disposal concepts are thus governed by the small waste volumes, the high initial radioactivity and heat generation, and the long isolation times. They can be subdivided into two groups, which we may call geology-based and technology-based, depending on whether the main obstacle to migration into the biosphere is determined by geological factors or not. Technology-based concepts will not be treated further in this book. They include the shooting of HLW canisters into space (extraterrestrial disposal), and the chemical separation of all transuranics from the waste (partitioning) and their conversion to stable isotopes by neutron bombardment (transmutation). Another technology-based concept is that of complete immobilization: the idea that it may be possible to convert the HLW into such a stable form that it would remain intact forever, no matter what geological situation it was deposited in. Although a guarantee of complete immobilization may never be possible, treatment of the wastes to produce a maximum of stability is part and parcel of all disposal strategies. It also has important earth science aspects. Immobilization techniques will thus be briefly reviewed below before going on to consider the main groups of geology-based disposal concepts.

A. Immobilization

Since containment is now generally recognized as a major aim in all radwaste disposal strategies, waste immobilization is an important prerequisite to disposal. This includes all artificial means of containment which can be attained by appropriate treatment of the wastes before emplacement in the repository. The main steps in waste treatment are as follows:

1. Extraction and concentration—involving processes such as distillation, calcination, and precipitation for liquid waste streams, and pulverization, compaction, and incineration of contaminated solids.
2. Fixation—involving incorporation as a mixture or in chemical combinations in an inert matrix, such as glass, ceramic, synthetic silicate minerals, concrete, or bitumen.
3. Encapsulation—involving enclosing the inert, waste-containing matrix (waste form) in a canister composed of one or several shells of inert metal, such as chrome–steel, lead, copper, or titanium, and/or other materials.
4. Buffering—involving emplacement of the waste canisters in an enclosing volume of fill material, such as bentonite, zeolite, soil or oxides of manganese, materials with special properties for protecting the integrity of the canisters and for retarding the migration of radioisotopes, should the other barriers be breached in the course of time.

The end result of all these processes is what is known as the waste package (Fig. 8).

The first principle of immobilization is solidification, and extensive research has been and is being carried out to determine the best way of transforming the large volumes of liquid HLW at present in interim storage into a solid waste form. From the point of view of industrial-scale processing, the emphasis is on simplicity (manipulation must be done behind metre-thick concrete walls, by remote control) and flexibility (the processes must stabilize all hazardous radioisotopes equally for materials of variable composition). From the point of view of transport, emphasis is on volume reduction and impact resistance (the risks associated with handling and transport accidents must be reduced to a minimum). And from the point of view of disposal, emphasis is on the resistance to leaching by water under a wide range of physical and chemical conditions (the radioisotopes must be trapped in materials of extremely low solubility). Unfortunately, these requirements are often in conflict and there is no consensus on the best strategy to be followed. Combine these uncertainties with the other unknowns (whether spent fuel will be reprocessed, what degree of partitioning is desirable) and it is clear that this preliminary step in any disposal strategy is at present in a state of flux. So geology-based disposal concepts must be discussed with a whole range of possible waste forms in mind, from spent fuel elements in canisters estimated to last only a few hundred years to wastes incorporated in synthetic silicate minerals whose natural equivalents have remained unaltered in many geological environments for a billion years or more.

It is the latter category of materials, synthetic rocks and minerals, which may eventually allow a strategy of complete immobilization to be followed. For instance, it has been shown that powdered metal oxides can be added to liquid HLW and the mixture treated by sintering and hot-pressing to give an aggregate of titanate minerals with all the major radioisotopes built into the crystal lattices (Figs. 14 and 59). These aggregates, called Synroc, and other similar ceramics (see Chapter 9, Section IV) have shown themselves to be much more resistant to leaching under pressure, temperature and water chemistry conditions similar to those expected in most repositories in comparison to the glasses which are currently favoured for HLW immobilization (see Chapter 8, Section IV). The synthetic rocks produced in this way contain minerals which occur naturally in many different geological situations, with

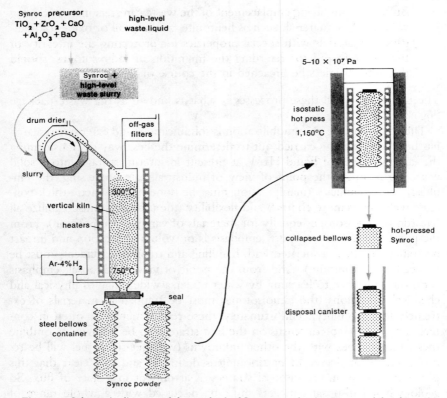

Fig. 14. Schematic diagram of the method of Synroc production suggested for the easier and more effective immobilization of HLW (see also Fig. 59). Typical phases in the final radwaste matrix are hollandite (containing Cs), zirconolite (containing Sr, actinides), and perovskite (containing Sr, lathanides, some actinides) (see Table IV). (From Ringwood, 1982, by permission of Sigma Xi, The Scientific Research Society.)

many different ages, so their use lends itself to assessing the long-term stability of such wastes. Even if complete immobilization cannot be guaranteed, the predictive capacity achieved by using such materials may give them the edge over wholly artificial materials for which there are no geological data.

B. Disposal in Deep-Mined Cavities

The disposal of HLW canisters in deep underground caverns has been the concept most consistently followed to date, and the one subjected to the most extensive research and development effort. It envisages a system of tunnels at depths down to 1500 m in the Earth's crust, with systems of drillholes in the tunnel floors designed to accommodate one or several waste canisters and an ample buffer of backfill material (Fig. 15). After the emplacement of the waste canisters was completed, the tunnel itself would be backfilled and sealed, as would the whole repository, including storage areas and access shafts, when it could no longer be extended or when there were no more wastes for dis-

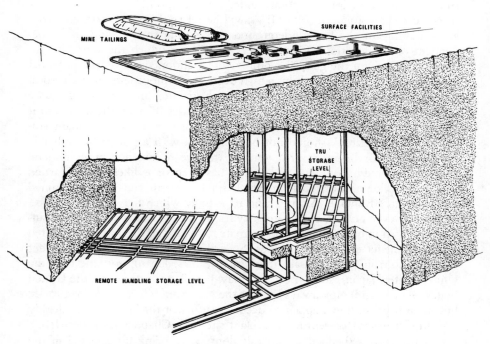

Fig. 15. Conceptual design of a deep-mined repository for both TRU wastes and HLW wastes (Waste Isolation Pilot Plant, New Mexico). See Fig. 50 for geological framework. (From Weart, in Fried, 1979, by permission of the American Chemical Society.)

posal. The ground surface above the repository would become a protected area in which future drilling activities or resource exploitation would be forbidden and in which a monitoring programme would be maintained as long as possible. However, since such societal commitment cannot be guaranteed for more than a few hundred years, only a small fraction of the required isolation time for most HLW categories, a geological system of predictable stability and with predictably very slow release mechanisms would be the main factor in site selection.

Since the main release mechanism will be circulating underground water, most concepts envisage a geological environment in which this is very slow and likely to remain so under the disturbing influences of site investigation, repository construction and waste emplacement. The whole problem is now the object of complex systems analysis, with extensive experimentation and mathematical modelling, assessment of failure scenarios and evaluation of the probabilities of future adverse processes and events, and predictions of the consequences of failure in terms of hazards and risks. Much of this work will be treated extensively Parts II and III of this book: the following is a brief summary of the various proposals under consideration.

Because the volume of solid HLW will be relatively small, most countries are searching for the best underground sites within their boundaries, independent of the location of the nuclear facilities at which the waste is produced. Apart from the general requirements mentioned above, the potential body of host rock should exhibit a high degree of homogeneity over a vertical distance of hundreds of metres and over a horizontal area of several square kilometres. There should be no large-scale fractures through the area, and it should lie away from earthquake zones and active or recently dormant volcanoes. The host rock should not be connected with a potentially valuable raw material, or be in an area where extensive drilling for other mineral resources (including oil, gas, and water) has been or is likely to be carried out. In addition, a host rock in which stable caverns can be easily mined would be an advantage. Although no rock body which fits all these requirements will ever be found, two rock types, rock salt and granite, occur sufficiently often in bodies which approximate to them that they are the favoured enclosing media in many countries. Several other are or have been under consideration, including basalt (the United States, Japan), volcanic tuff (the United States, Japan), shale (Belgium, Italy, Japan), and anhydrite (the United States, Switzerland), not always because they are *a priori* favourable host rocks, but often because they happen to occur in sufficient thickness below already operational reactor or reprocessing plant sites (see Chapter 8, Section III).

Salt formations (bedded salt or salt domes) are being investigated in the United States, West Germany, Denmark, Holland, and Spain, and underground testing under simulated repository conditions has been carried out in the Asse II salt mine (see Fig. 12) and is underway in the United States at the Waste Isolation Pilot Plant, New Mexico (see Fig. 50). In spite of several negative characteristics (e.g., solubility and evidence for dissolution, brine content, heterogeneity, and association with valuable resources), salt as a potential radwaste host rock remains a leading choice in those countries in which salt bodies of sufficient size exist. In other countries, and as an alternative to salt in the United States, granite and related coarse-grained silicate rocks are under investigation (Canada, Sweden, Switzerland, Finland, France, Japan, and the United Kingdom), although in one respect—the ubiquitous presence of cracks and concomitant potential water circulation paths—it does not show the required properties. Test facilities are at present in operation in the United States (Climax Mine, Nevada Test Site), Sweden (Stripa Project) (see Fig. 62), and Switzerland (Grimsel rock laboratory) and will shortly be opened in Canada (see Fig. 87).

C. Disposal in Very Deep Boreholes

An alternative technique for emplacing HLW deep in the earth's crust would be to use the lower parts of very deep boreholes. Deep-mined cavities can be constructed down to ∼ 1000 m but below this they become technically difficult and prohibitively expensive. This means that they will always lie within the potential zone of circulating groundwater. In contrast, boreholes can be drilled down to at least 10 km and there is evidence that many rocks at this depth (corresponding to temperatures of 300° C and pressures of 4 kbar) no longer contain interconnecting pore spaces or cracks, the prerequisite for water circulation. With an extension of present-day rotary and "big-hole" drilling techniques, hole diameters of ∼ 1 m may be attainable to depths of 5 km, and diameters of 30 cm down to 10 km or more. With present-day projected HLW cylinder sizes (32–71 cm diameter, according to country) this would permit emplacement and buffering at depths of 6–8 km in crystalline rocks.

Various countries have carried out studies to assess the feasibility of this type of disposal procedure (the United States, Denmark, Switzerland and Australia), although no testing related to radwaste disposal has been carried out. There is agreement that it is a practical proposition with present technology and that it would have definite advantages from a sociopolitical point of view (obviously increased safety, easily grasped procedures, and possibility of decentralizing disposal facilities). The main difficulty is that its adoption would necessitate the abandonment of the only immobilization process which has been developed to an industrial production stage over the past 20 years and which is now the official procedure in the United States and France: vitrification of HLW as a borosilicate glass. Most deep borehole concepts envisage emplacement at depths of 2–10 km (depending on enclosing medium), corresponding to ambient temperatures of 60–300° C. Even if kept in storage for 40 years, it is calculated that glass temperatures would exceed 200° C in many situations and, if water were present in only small amounts, would rapidly devitrify and disintegrate (see Chapter 8, Section IV,B). Thus, for effective immobilization as a preliminary to very deep borehole disposal, a ceramic waste form would have to be used, at least for HLW from fuel reprocessing. It may be that suitably encapsulated spent fuel could be disposed of directly by this method, but then, of course, up to 20 times the number of very deep boreholes would have to be sunk.

D. Deep Underground Melting

Whereas the very deep borehole concept spaces the HLW canisters deliberately far apart to minimize the temperature buildup in the hole, the deep underground melt process (DUMP) is conceived as the result of doing the opposite: all the HLW would be concentrated at one place, where it heats itself up to melting point, together with a large volume of the surrounding rock (Fig. 16). The original concept envisaged deep shafts or boreholes ending in a void with a volume of about 5000 m^3 at a depth of 2–5 km (excavated by an underground nuclear explosion or by advanced mining techniques). A favourable host rock would be a non-carbonate-bearing shale, fine-grained and homogeneous over a thickness of 300–400 m and impervious to water, or a granite. HLW in liquid or solid form would be inserted in the void rapidly (or slowly under constant cooling) until the void was filled. The heat generated by the wastes would then accumulate, since rocks are very poor heat conductors, and eventually temperatures of >1000° C would be attained, melting the surrounding rock and dissolving the radionuclides in a growing sphere of molten material. Growth would continue until such time that the heat conduction outwards balanced the radioactive heat production (in the model calculations, after ∼ 90 years, resulting in a maximum radius of melting of 80–100 m). From this time on, slow cooling would take place, with complete crystallization of the melt (and immobilization of the radionuclides) only after several hundred years. During the whole of this time, the cooling mass would be protected from any possible water incursion by a thermal screen, the enveloping surface at which water would vaporize, which is calculated to lie outside the margins of the original molten sphere throughout this period. After complete crystallization and cooling (after 1000 years), it is estimated that the relative hazard due to the long-lived radioisotopes in the now strongly "diluted" waste (volume waste:volume molten rock = 1:1000) would be ∼ 1% of that of the naturally occurring uranium mineral, pitchblende.

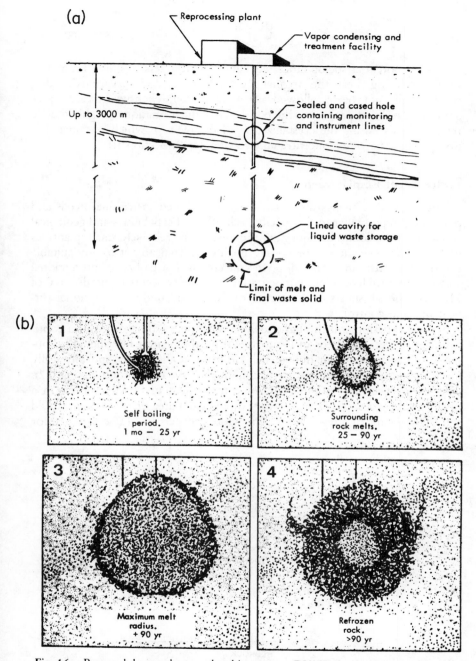

Fig. 16. Proposed deep underground melting process (DUMP) for the disposal of HLW. (a) Liquid waste emplacement in a mined cavity in the bedrock below the reprocessing plant. (b) Four stages in the evolution of the cavity: (1) period of filling and self-boiling (up to 25 years); (2) start of melting after closure (25–90 years); (3) attainment of maximum melt radius; and (4) gradual refreezing of molten rock and waste mass. (From Gilmore, 1977, by permission of Noyes Data Corporation.)

Despite potential sociopolitical advantages (cheapness, disposal located at waste-producing sites), the original DUMP concept and its various modifications have not been followed up with a comprehensive research programme. The main problem is the possibility of the formation of highly mobile residual fluids during the crystallization of the melt, in which certain radioisotopes may become concentrated (see Chapter 9, Section III,A).

E. Ice Sheet Emplacement

The disposal strategies discussed so far are tailored to national needs and potentials, in particular to the specific technological capabilities and geological situation of the country in question. Because of the relatively small quantities of HLW, however, it is theoretically possible to transport it to any suitable site on the earth, and some disposal concepts take a global or international view of the problem. One representative of these concepts is the disposal of HLW in the ice sheets of Antarctica or Greenland under international surveillance and control.

Several variations on the basic idea of isolation at some depth in a continental ice sheet, hundreds of kilometres from the nearest biotope and thousands of kilometres from human habitation, in a chemically and physically inactive environment with no circulating fluids, have been proposed (Fig. 17). The most recent suggestion envisages emplacement of HLW canisters in a small, specially designated area of the Antarctic (an area of ~ 700 km² would suffice) at depths of a few metres. The waste canisters would be designed for

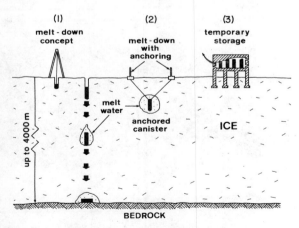

Fig. 17. Various concepts for the disposal of heat-producing radwastes in the Antarctic or Greenland ice sheet: (1) Emplacement in boreholes, followed by melt-down; (2) shallow emplacement with melt-down prevented by surface anchors; and (3) above-surface repository.

self-shielding during transportation and emplacement (e.g., a 10-cm-thick lead container would be sufficient to shield reprocessed HLW after some years of interim storage). The size, radioisotope content, storage time, and spacing would be arranged to ensure that the thermal processes only allowed a melt-down through the ice of the order of 100 m (a condition which could be guaranteed with an appropriate anchoring system). Using these and other precautions, it is estimated that even a catastrophic climatic change resulting in rapid melting and/or flow instability (surging) of the ice could not cause the canister to reach the biosphere in the next few thousand years. However, most glaciologists agree that prediction into a more distant future is hardly possible for such a geologically unstable body as a mass of ice (see Chapter 13), so the ice sheet emplacement concept would have to be coupled with reprocessing and with partitioning of the long-lived α-emitters, which require much longer isolation times.

The Antarctic Treaty of 1959 (due to be renewed in 1989) specifically prohibits the disposal of radioactive waste on the continent, a clause which can only be lifted by agreement among all the treaty nations (Art. V, 1 and 2). Directed research will thus only be possible if it becomes clear that ter-restrial options within national boundaries are not acceptable, and if the op-position of the nonnuclear nations can be overcome.

F. Subseabed Emplacement

Another class of disposal concepts for HLW which involve international cooperation and long-distance transport to sites remote from habitation and low in known release mechanisms are those based on emplacement within the sediments below the seabed. Some earlier ideas, such as disposal in deep sea trenches or in mid-oceanic fracture zones, have now been abandoned, but the concept of emplacement of HLW canisters in deep-sea clay layers in the middle of stable areas of the ocean floor has been gaining momentum since 1973 (see Chapter 12 and Chapter 15, Section III). Recently, research and development efforts in this direction have been placed on a firm inter-national footing with the establishment of the Seabed Working Group (par-ticipating countries: the United States, the United Kingdom, France, Canada, Japan, Holland, West Germany and Switzerland, under the auspices of the OECD Nuclear Energy Agency and in cooperation with the Commission of European Communities). Although the London Convention of 1977, of which the participating countries are signatories, prohibits the dumping of HLW into ocean waters or onto the ocean floor, it is hoped to gain inter-national recognition for controlled emplacement under the seabed, if the re-sults show that the concept is technically feasible.

Present research is geared to a reference concept (Fig. 18) which will be

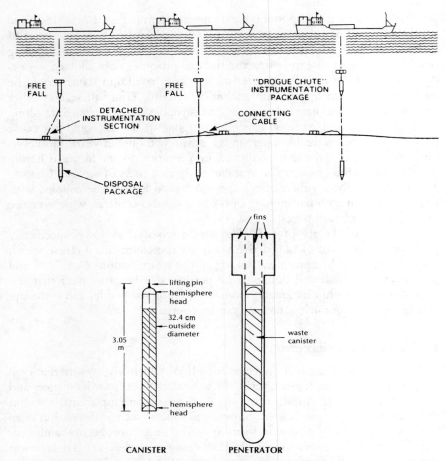

Fig. 18. Emplacement concept of the international Subseabed Disposal Program. Recent tests have confirmed that emplacement at depths of up to 30 m in deep-sea clays (at water depths of 4000 to 6000 m) using such techniques is feasible. (From Kelly and Shea, 1982, by permission of Woods Hole Oceanographic Institution.)

continuously modified in the light of future results (see Chapter 15, Section III). The waste, vitrified HLW or encapsuled spent fuel rods, will be stored and then packaged in 3-m-long containers of 30 cm diameter. The upper limit for thermal output at the time of transportation and emplacement will be 1.5 kw per canister. The canister materials and thickness will be chosen to give containment times of 500 yr in an oxygenated, high-pressure seawater environment, although total sediment enclosure (a much less corrosive environment) is the aim. The reference site is in the North Pacific, in the middle of a 10^8 km^2 area of the seafloor which is covered with 50–100 m of red clay and 5000–6000 m of water. The region is known to have been extremely stable, uniformly sedimented, and untouched by large climatic variations for a period of at least 10 m.y. and is expected to remain in the same condition far into the future. The canisters would be inserted individually by penetrometer (free-falling, finned projectiles) or by deep sea drilling at depths of 30 m within the clay and hundreds of metres apart. Experiments and computer modelling have shown that the thermal perturbation around such a canister would be slight (maximum radius of the 100° C isotherm = 80 cm) and that transuranics like ^{239}Pu will not migrate more than a few metres from a breached canister in 100,000 years. From many other points of view, including retrievability, cost, avoidance of natural resources, and predictability of future events, the concept is judged to be at least as viable as the land-based strategies being followed as top priority by most countries. But will it ever be acceptable within the framework of international politics and the Law of the Sea?

V. CONCLUDING REMARKS

In this chapter, we have surveyed the different methods, in use, under development, or proposed on paper, of radwaste disposal. All the options and alternatives discussed hold out some promise of success if developed and carried out with modern technological and institutional control. This statement can be made because, slowly, proper note is being taken of past negligence and its sometimes catastrophic results. Not attempted here was an assessment of the relative advantages and disadvantages of the various strategies. This would involve many technological and political aspects which lie outside the scope of this book. It is clear, however, that each alternative must be placed within a geological frame of reference, which will be an important element of any final assessment. Long-term confinement or isolation will depend on the effects of natural processes going on at the Earth's surface, at the bottom of the ocean, and within the upper few kilometres of the Earth's crust. These processes are the theme of Part II of this book, in which the various disposal concepts are viewed from an earth science perspective.

SELECTED LITERATURE

General reviews and sources: U.S. Energy Research Development Administration (ERDA), 1976 (Vol. 4); Gilmore, 1977; Organization for Economic Cooperation and Development (OECD), 1977; 1982a, b; Macbeth *et al.*, 1978; Jet Propulsion Laboratory, 1978; U.S. Department of Energy (DOE), 1979c; Roberts, 1979; Harmon *et al.*, 1980; Fyfe, 1980;

Milnes *et al.*, 1980; Spoljaric, 1981; Lindblom and Gnirk, 1982; Donath and Pohl, 1982; Cope *et al.*, 1983.

Liquid LLW–ground percolation: Hubbell and Glen, 1973; ERDA, 1975; Price and Ames, 1975; Anderson, 1975; Gilmore, 1977 (pp. 258–280); Lipschutz, 1980 (pp. 129–132); Kasper, 1981; Barraclough *et al.*, 1982; Last, 1983; Olsen *et al.*, 1983.

Liquid LLW–deep well injection: Adam *et al.*, 1974; Gilmore, 1977 (pp. 208–212); Spitsyn *et al.*, 1977; Waller *et al.*, 1978; Brown *et al.*, 1979; Brawner, 1980 (pp. 507–520); Smith, 1981; Perkins, 1982.

Liquid LLW–grout injection: Lutze, 1982 (pp. 849–888); Perkins, 1982; Sun, 1982; International Atomic Energy Agency (IAEA), 1983b; Stinton *et al.*, 1983.

Solid LLW–controlled dumping at sea: Bowen and Livingston, 1978; Deese, 1978 (pp. 45–58); IAEA, 1978a, b; Colombo *et al.*, 1979; OECD, 1979a, 1980a; Templeton, 1980; Jackson, 1981 (pp. 17–54).

Solid LLW–shallow land burial: National Academy of Sciences (NAS), 1976; Gilmore, 1977 (pp. 201–207); ERDA, 1977; Carter *et al.*, 1979 (pp. 731–918); IAEA, 1980b (Vol. 2, pp. 253–269); State Planning Council (SPC), 1981; Coates, 1981 (pp. 635–670); Burton *et al.*, 1982.

Solid LLW–disposal in mined cavities: National Cooperative for Radioactive Waste Storage (NAGRA), 1981; Winograd, 1981; Burton, 1982; Herrmann, 1983 (pp. 159–173).

Management of uranium mill tailings: OECD, 1978; DOE, 1979a; Argall, 1979 (pp. 109–124, 471–486); Brawner, 1980; IAEA, 1982a.

HLW and LLW–immobilization: Gilmore, 1977 (pp. 50–103); Becker *et al.*, 1979; Kibbey and Godbee, 1980; Butlin and Hills, 1982; DOE, 1982a; Lutze, 1982 (pp. 299–308); Ringwood, 1982; Roy, 1982; IAEA, 1983c; NAS, 1983 (pp. 43–109).

HLW–disposal in deep mined cavities: Fried, 1979 (pp. 13–36); Carpenter and Martin, 1980; IAEA, 1980b; Ahlström *et al.*, 1981; Gonzales, 1982; Swedish Nuclear Fuel Supply Co. (KBS), 1983; NAGRA, 1983a.

HLW–disposal in very deep boreholes: O'Brien *et al.*, 1979; Forex Neptune, 1980; Ringwood, 1980; Elsam/Elkraft, 1981.

HLW–deep underground melting: Schwartz *et al.*, 1976; Gilmore, 1977 (pp. 180–188); McCarthy, 1979 (pp. 261–264); Heuze, 1982.

HLW–ice sheet emplacement: Zeller and Saunders, 1972; Bull, 1975; Robin, 1975; ERDA, 1976 (pp. 25.52–25.61); Philberth, 1977; Macbeth *et al.*, 1978.

HLW–sub-seabed emplacement: Bostrom and Sherif, 1970; Turekian and Rona, 1977; Deese, 1978; Anderson, 1979; Hollister *et al.*, 1981; Jackson, 1981 (pp. 55–78); Kelly and Shea, 1982; Subseabed Disposal Program (SDP), 1983.

PART II

EARTH SCIENCE PERSPECTIVES

The main part of this book concerns the contribution of the earth sciences to an understanding of the long-term performance of any radwaste isolation system. In this respect, the geological data base contains three types of relevant information. It contains information on the various types of enclosing media, both repository host rocks and radwaste matrices, and on the processes leading to their formation in and on the Earth's crust. It contains information of relevance to assessing the possible release mechanisms which might affect the integrity of a surface or subsurface repository. And it contains information on the many different types of natural analogues of repository-like situations which have existed in the past and which are the main basis for very long-term assessments. These three types of data are closely interconnected and, to emphasize interrelationships rather than compartmentation, we shall use a subdivision based on geological environment.

After introductory chapters on the Earth's crust (Chapter 3) and geological time (Chapter 4), we first deal with the surface environment and the natural processes active within the skin of loose, weathered materials which cover most of the bedrock (external processes). These can be studied in action at the present time by direct observation and they are often intuitively understood or unconsciously registered by people in everyday life, since they are an integral part of man's natural environment. Nevertheless, many such processes work so slowly that they are mainly known through their effects rather than through experience of their action. This is particularly true of the biogeochemical processes described in Chapter 5, of which the most important are weathering and soil formation. Site selection and safety assessment of surface deposits of solid LLW and uranium mill tailings depends to a large extent on a deep understanding of natural leaching processes, of water flow patterns, and of pollutant migration mechanisms within the zone of weathering. The subject matter of Chapter 6, denudation and deposition, is of lesser direct relevance, since sites at which such processes are known to be important (e.g., areas of high relief, flood plains) are generally avoided for other reasons, but it provides an important link to Chapter 7, which describes the formation and properties of sedimentary rocks. Some sedimentary rocks are consolidated land deposits, but most of the sedimentary formations envisaged as host rocks for deep-mined repositories (bedded salt, shale, etc.) were originally deposited in shallow seas. An understanding of the process of transformation of soft marine sediment to consolidated rock leads to a better appreciation of the properties, structure, and long-term behaviour of these materials. A similar philosophy is also followed in the succeeding chapters on volcanic rocks (Chapter 8) and crystalline rocks (Chapter 9), but these are somewhat more far ranging, including sections on volcanism as a potential disruptive process for any repository type and on volcanic glass, metamorphic rocks, and minerals as natural analogues for the various vitreous and ceramic waste forms.

Chapters 7–9 deal essentially with processes which have taken place within the Earth's crust and which have resulted in the formation of the different types of rock (internal processes). The next two chapters focus on the present-day subsurface environment, at those depths which are relevant for the construction of deep-mined repositories. Chapter 10 treats the physical aspects (temperature, pressure, and fluid flow) and the long-term effects of repository construction and radwaste emplacement on the different parameters, whereas Chapter 11 discusses the geochemistry of rock–fluid and rock–fluid–waste interaction, particularly with respect to the leaching and migration of radionuclides underground.

The final chapters in Part II do not follow logically from the above sequence. Chapter 12 is concerned with ocean processes and provides a background for assessing the practices of releasing liquid and solid low-level wastes to the ocean waters and the proposals for emplacing solid high-level waste in the sediments of the ocean floor. Chapter 13 addresses the question of the possibility of a future continental glaciation, the main event likely to affect the long-term stability of any repository, and summarizes the geological evidence relating to past and future climatic change. Part II is intended as a summary of those aspects of earth science which are of relevance to the problem at hand: the application of this data to long-term prediction and site selection is treated in Part III.

CHAPTER 3

The Earth's Crust

I. INTRODUCTION

From the foregoing chapters it is clear that judgements as to whether the risks associated with the proposed radwaste disposal methods on and within the Earth's crust are acceptable or not depend to some extent on a knowledge and understanding of crustal processes. For instance, the concept of emplacement of HLW in mined repositories in deep, dry geological environments depends not only on finding such a favourable location but also on proving that the location is likely to remain favourable over the required isolation time. In other words, the concept presupposes a detailed knowledge of the Earth's crust at depth and an advanced understanding of processes which have affected the crust in the past and will continue affecting it in the future. Because of the variety of different concepts, this backcloth to the problem of radwaste disposal encompasses practically the whole of the science of geology, defined as the study of the structure and history of the solid Earth, and overlaps considerably into other areas of earth science, such as hydrology, oceanography, and meteorology.

The aim of this chapter is to survey the processes which have contributed to the shaping of the earth's crust as we see it today, as a background to more detailed consideration of those most relevant to the radwaste problem in later chapters. First, we consider the structure of the interior of the Earth, focussing particularly on its outermost shell, the crust. From this static picture we move to the present-day dynamics of the earth (earthquakes, volcanic activity, and heat flow) and find that the evolution of the crust can be understood in terms of a dynamic model of the Earth known as global plate tectonics. The materials of the crust, the rocks themselves, all tell a story of individual development which, when pieced together, leads to a remarkably coherent picture of the evolution of the two main types of crust, oceanic and continental. The chapter ends by introducing water as a major factor in crustal evolution and as the main agent of radionuclide transport in the Earth's surface environment.

II. INTERNAL STRUCTURE OF THE EARTH

The constitution of the Earth's interior has drawn the interest of philosophers since ancient times. Very early it was realized (mainly from observations of volcanic activity) that the inside of the Earth must be very hot. This was confirmed during the Middle Ages by the experience of temperature increasing with depth in mines. The general concept of the Earth having a hot, molten core and a cold, solid crust, first formulated by Descartes in 1644, then gradually emerged. With the advent of the theory of gravity, followed by the development of methods for determining the gravitational constant in the eighteenth century, it became clear that the average density of the Earth was much higher than the density of most materials exposed at the surface. Slowly, the accumulated evidence from more accurate surveying, from the gravity anomalies discovered around mountain chains, from observations on large earthquakes, and from the study of meteorites led, in the nineteenth century, to a picture of the Earth's interior remarkably similar to that of the present day. The French scientist Daubrée, for instance, suggested in 1868 that the Earth consisted of a very dense, liquid core (thought to consist of an iron-nickel alloy with a density of ~ 8 g/cm^3, similar to iron meteorites), a less dense, solid "stony" layer (thought to be made up mainly of the mineral olivine, which has a density of 3–4 g/cm^3, similar to stony meteorites) and a light, solid crust (made up of rocks similar to those exposed today at the Earth's surface). At the same time, ideas on the internal temperature distribution were developing, based on the hypothesis, propagated rigorously by the English physicist, Lord Kelvin, that the Earth originated as a sphere of molten rock and reached its present state by slow cooling and solidification.

This was the status of the model at the end of the last century. Three main developments have allowed the modification and refinement of this picture to its present form (Fig. 19). First, the development of the science of seismology, the quantitative study of the propagation of earthquake vibrations,

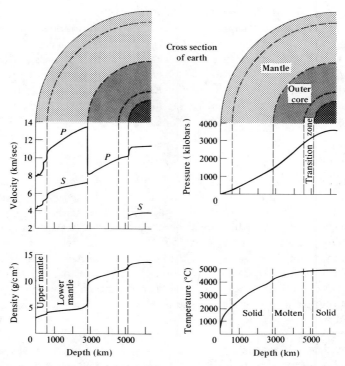

Fig. 19. Summary curves showing the average variation of seismic velocities, density, pressure, and temperature inside the earth. (From Bolt, 1982, by permission of W. H. Freeman and Co.)

provided a means of "X-raying" the inside of the Earth. This started with the establishment of a worldwide network of earthquake recording stations (seismographs) between 1890 and 1900, and has continued ever since, concomitant with improvements in instrumentation and seismograph distribution. Second, the growth of the discipline of geothermics contributed greatly to our knowledge of conditions in the Earth's interior and has shown that heat transfer processes lie behind most geological phenomena. This was stimulated by the discovery of radioactivity and, between 1900 and 1910, by the realization that the heat produced by the radioactive decay of radioisotopes present in small amounts in all rocks, soils and waters contributes significantly to the heat budget of the Earth (accounting for 40–80% of the surface heat flow). Third, techniques were developed for testing the properties of materials under pressure and temperature conditions analogous to those at great depth in the Earth. This area of experimental petrology joined forces in the last three decades with the study of meteorites and volcanic rocks to provide data on the chemical and mineralogical constitution of the Earth's interior. As we shall see later, all three disciplines—seismology, geothermics, and experimental petrology—are also deeply involved in the radwaste problem in ways unrelated to their contributions to our understanding of the "bowels of the Earth." Although many other types of data have been integrated into the current model of the internal structure of the Earth, including astronomical, gravitational, magnetic, and shockwave experimental, we concentrate on data from these three areas in the following.

A. Seismological Data

An earthquake (or an underground nuclear explosion) involves a sudden release of energy. Part of this takes the form of elastic waves which travel through the Earth. These waves are subdivided into body waves (radiating from the focus of the earthquake at depth) and surface waves (radiating from the epicentre, a point on the Earth's surface immediately above the focus, and confined to the surface layers). Although the surface waves are the ones causing the earthquake damage, the body waves are those which enable us to "see" into the Earth's interior. The faster of the body waves (P, primary) are longitudinal waves similar to sound waves and can travel through both solid and liquid materials. The slower of the body waves (S, secondary) are transverse waves similar to light; they can be polarized, but they are not transmitted by all materials. For the S waves, the opaque materials are those with zero rigidity, i.e., liquids. Both P and S waves travel with different speeds in different materials, depending on the elastic constants and density. Refraction and/or reflection take place where these material properties change. Analysis of the travel times of the different reflected and refracted waves between the focus and the recording stations distributed over the Earth's surface enable the changes in material properties in the Earth's interior to be mapped.

In spite of the complexity of the computations, there is now general agreement on the main outlines of the picture. There are three main discontinuities in the body of the Earth. The shallowest one is the most important from our point of view. It marks the base of the crust and is known as the Mohorovicic discontinuity or the Moho. It is a good seismic reflector, is characterized by a sudden increase in P-wave velocity (v_P), and represents a compositional change from light to dense material (see Fig. 20). Its depth varies from 5 to 60 km. Under the oceans it generally lies ~ 10 km below the sea floor, whereas in continental areas away from the main mountain chains it is usually between 30 and 40 km deep. The next deepest discontinuity, the Gutenberg discontinuity (discovered by B. Gutenberg in 1912), lies at ~ 2900 km depth and defines the boundary between mantle and core.

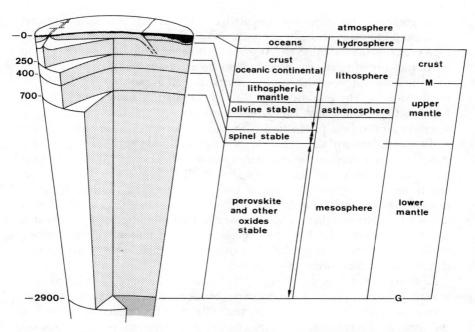

Fig. 20. Diagrammatic segment of the earth's crust and mantle showing the different types of layering. Depths (km) on left. M, Moho; G, Gutenberg discontinuity. (From Smith, 1982, by permission of Trewin Copplestone Books Ltd.)

At this boundary, *P*-wave velocities are sharply reduced, indicating a change in composition, and the *S* waves are stopped altogether, indicating a change from solid to liquid. The third discontinuity, at a depth of 5200 km, is a relatively recent discovery. It is thought to mark a change in state from liquid to solid, but only a slight change in composition. Recent refinements in methodology have enabled further subdivisions to be made. Of particular importance for the discussion of Earth dynamics (see Chapter 3, Section III) is the subdivision of the upper mantle on the basis of variations in *S*-wave velocity. The marked reduction in v_S at a depth of 80–100 km marks the boundary between a solid, strong upper part (lithosphere) and a partially liquid, weak zone (asthenosphere) in the upper mantle. It is probably not a change in composition, since it is not reflected by a corresponding change in v_P.

B. Geothermal Data

The present temperature distribution in the Earth's interior is very difficult to deduce and can only be estimated by combining often circumstantial evidence from many different areas. Geothermal data provide two important fixed points in this otherwise speculative field: surface heat flow measurements and estimates of the heat production due to radioactive decay in common crustal rocks and minerals. At present, there are more than 5000 reported measurements of terrestrial heat flow giving a average rate of 53 W/m^2 for the continents (30% of the measurements) and an average rate of 62 W/m^2 for the oceans. The total heat loss through the Earth's surface is about 10^{21} J/year (cf. present global energy consumption of 3×10^{20} J/year). Measurement of the heat production of surface rocks depends on analyzing the rocks themselves for their content of long-lived radioisotopes. Combining these with estimates of the amounts of different rocks in the Earth's crust give heat production values of around 4×10^{20} J/year. The difference between this and the heat loss estimate gives the contribution of heat from the mantle to the crustal heat budget. Other estimates of the amount of heat flowing from the mantle into the crust are derived from the linear relationship between heat flow and heat production in continental areas; extrapolating this line to the zero heat production axis reveals an input from nonradiogenic sources of the same order of magnitude.

Analysis of local, regional and global data of this kind, combined with knowledge of the thermal conductivity of rocks, allows temperature–depth curves, or geotherms, to be constructed with some confidence to the base of the crust. Moho temperatures vary considerably, depending on the thickness and type of crust above, but generally fall in the range 400–800° C. Below the Moho, temperature estimates are more speculative (Fig. 19) and are based on syntheses of seismological and geothermal data with the results of experimental geochemistry.

C. Data from Experimental Petrology

At the present time, machines are available which allow the physical and chemical behaviour of materials to be studied at pressures up to 300 kbar (equivalent to depths of 650 km) and temperatures up to 1000°C. All the evidence points to the mineral olivine being the main constituent of the mantle immediately below the Moho. Since the pressure gradient in the upper mantle is known, the interpretation of the zone of low v_S (asthenosphere) as one with small quantities of melt can be tested experimentally by determining the melting point of olivine at high pressures and temperatures. The melting temperature also places an upper limit on the temperature at any depth in the parts of the mantle known to be solid. Other experiments allow the pressure-temperature (PT) conditions of significant phase changes to be determined. For instance, at very high pressures, equivalent to depths of 400 km, and at appropriate temperatures, olivine undergoes a phase change to a more compact crystal structure (spinel structure) (Fig. 20). This seems to correspond to the marked zone of sharply increasing v_S at ~ 400 km depth, discovered earlier by the seismologists. Further down, at ~ 700 km, a second increase in v_S may be related to a further phase change, from spinel to an even denser structure similar to that of the mineral perovskite (see Chapter 9, Section IV) and magnesium oxide.

The present picture of the Earth's interior, however, does not depend on the application of a few specific methods, but rather on a continual sifting of often very indirect evidence from all areas of science. The above discussion is of necessity incomplete and the picture presented essentially static. In the following sections, we now single out the Earth's crust for more detailed consideration and review the evidence that we are dealing with a dynamically evolving layer which has had a long and complicated history.

III. GLOBAL PLATE TECTONICS

The evidence that the Earth's surface is in continual, although often extremely slow, motion is indirect, and the idea of a dynamic Earth was a comparative latecomer in the history of scientific thought. Earthquakes and volcanic eruptions were long considered local phenomena without a pattern. The remains of marine organisms on high mountain peaks, and the folded and dislocated rock layers in mountain chains like the Alps, were long explained by Noah's flood, later by the slow contraction of the Earth, raising mountain chains like wrinkles in the skin of a drying apple. The birth of geology as a science at the end of the eighteenth century is marked by the general acceptance among geologists that many features of the Earth's crust can only be explained by general mobility—uplift, subsidence, displacement, tilting, fracturing, and intrusion. But battles continued throughout the nineteenth century on several fronts: how long a time scale, how continuous the movements, how large the displacements, how big the forces? The controversy around the size of the displacements, whether the continents and oceans had always remained more or less in their present position or whether the continents had drifted apart and together, and oceans opened and closed, continued up to the 1960s. Then we lived through a veritable revolution in geologic thinking, with the development of a new consensus in favour of an Earth model dominated by large-scale and continuous lateral displacements of lithospheric masses, or "plates." We shall look at the evidence for this theory, which is known as global plate tectonics, from three different points of view: the present-day patterns of activity, the indications of large-scale crustal movements in the past, and the geological corollaries of such movements at the boundaries between the plates. Global plate tectonics, as a unifying theory of the Earth, has become the starting point for all attempts at predicting future events, which is the essence of the radioactive waste disposal problem.

A. Present-Day Mobile Zones

The first group of observations which led to the idea of plate tectonics are those concerned with present-day activity, particularly the present distribution of earthquakes, volcanoes, and anomalously high heat flow (Fig. 21). By the 1960s, it had become quite clear that these were confined to a network of narrow mobile zones across the Earth's crust, separating large segments of crust within which seismic activity and volcanism were rare. This led directly to the concept of plates, stable or rigid regions which cover the Earth's surface like a mosaic but which are in constant motion relative to each other. The mobile zones represent zones where two plates move towards, past, or away from each other (convergent, transform and divergent plate boundaries, respectively). Continued study of the depth distribution and type of earthquakes, the types of volcanoes and volcanic products, and the heat flow distribution confirmed this qualitative picture and also suggested that the lithospheric plates were thin relative to their areal extent (~ 100 km thick) and that their movement was facilitated by the weakness or low viscosity of the underlying asthenosphere (see Fig. 20). In addition, the mass of data on continental geology which had accumulated up to this time showed that the rocks formed during the youngest period of earth history (see Chapter 4, Section III) were fractured, squeezed, or otherwise disturbed in the region of the present-day mobile zones, often forming the high mountain chains of the world, whereas rocks of the same age within the plates were relatively undisturbed. The signs that the plate boundaries had remained active over long periods of time were unmistakable and in agreement with the second group of observations, the indications of past movement.

B. Continental Drift and Seafloor Spreading

Our knowledge of the way in which the plates have moved, in which direction and by how much, we owe to two newcomers on the earth science scene: the study of ancient magnetic fields and the study of the ocean floor. Early in this century, it became clear that the jigsaw-puzzle fit of certain continents (South America and Africa, Australia and Antarctica) was not only a matter of the shape of the coastlines or continental shelves but also of the correspondence of the geological zones within the continents. Nevertheless, except for some prominent exponents, few people took the idea of continental drift, the theory that the continental fragments had drifted apart, often thousands of kilometres, in the geologically recent past, very seriously. In

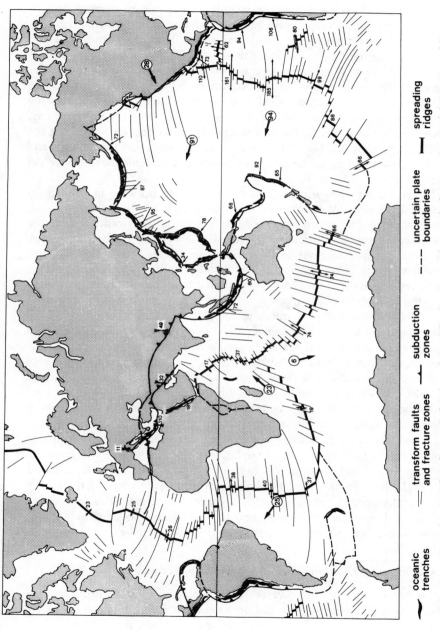

Fig. 21. Map of present-day plates and plate boundaries. The directions and rates of relative motion (in mm/year) are shown at selected points on the plate boundaries, together with the computed motions of the plates relative to six selected hot spots (in mm/year). (Data from Smith, 1982 and Bott, 1982.)

oceanic trenches

transform faults and fracture zones

subduction zones

uncertain plate boundaries

spreading ridges

postwar years, however, it became technically possible to measure the remanent magnetism of many rocks and to determine the position of the magnetic poles at the time of the rock's formation. Since age dating had also reached a measure of sophistication, this allowed the definition of "polar wandering paths" for each continent, that is, the variation in position of the Earth's magnetic poles with respect to that continent far back into Earth history. This study of ancient magnetic fields (paleomagnetism) showed, first, that the drift hypothesis was, in fact, true (each continent showed quite different magnetic poles at some specific time in the past, due to displacement since the ancient magnetism had been "frozen in") and, second, that the amount and speed of movement could be estimated (average displacement rates were determined to be $\sim$ 1 cm/year).

The second set of new data came from the great upswing in research in oceanography after World War II. The bathymetry of the oceans was surveyed in detail, the ocean floor sediments were dredged and drilled, the oceanic crust was scrutinized with natural and artificial earthquake waves, and the magnetic variations in the oceanic areas were studied, to mention but a few of the major advances. The result was the hypothesis of ocean-floor spreading in the early 1960s, the idea that new oceanic crust is generated along the mid-ocean ridges (see Fig. 23) and slowly gets dragged away from the ridges to make room for more as the plates on either side move apart. The successive additions of material can be mapped and dated, giving rather precise movement pictures for the opening of several oceanic areas, particularly of the Atlantic and the eastern Pacific, and allowing the positions of continents and oceans to be reconstructed at different points in Earth history (Fig. 22). The divergent boundaries at which new ocean floor is created are often called constructive; their counterparts are the convergent boundaries at which ocean floor is destroyed or consumed.

It is thought that there must be an approximate balance between construction and destruction since there are indications that the surface area (volume) of the Earth has not changed much since its origin. The ocean floor seems to be underthrusting other lithospheric plates around the whole of the Pacific Ocean at the present time (see Fig. 21), causing the circum-Pacific deep-sea trenches and volcanic arc systems and the increasing focal depth of earthquakes on passing towards the continents from the trenches.

C. Geological Corollaries of Plate Boundary Activity

This rather oversimplified picture of global plate tectonics as deduced from the geophysical, volcanological, and oceanographic evidence, is supported by the geological history of the present land areas over the past 200 m.y. This time span is, of course, far in excess of that required for the discussion of

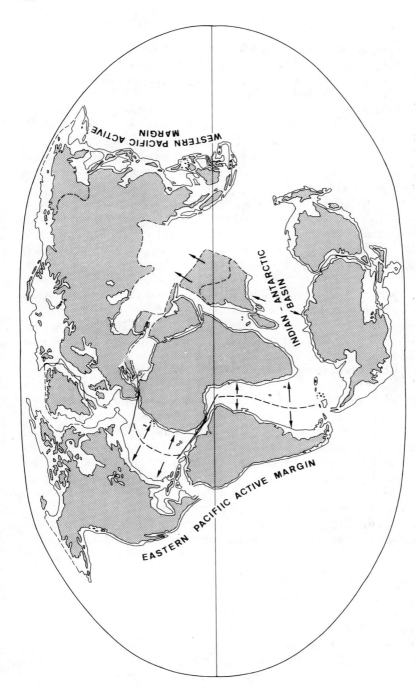

Fig. 22. World map showing the estimated relative positions of present-day land masses (stipples) and shallow shelf seas about 90 m.y. ago (compare with Fig. 21 to see the long-term effects of continental drift and sea-floor spreading). (From Owen, 1983, by permission of Cambridge University Press.)

radwaste disposal, but it is reasonable to assume that the picture gives a framework for discussing the geologically immediate future—say, the next 1 m.y. During this time, for instance, America will have moved another 2 km away from Europe. Kilometre-wide strips of new oceanic crust will have developed at various places on the ocean floor by molten rock rising and cooling in contact with sea water. Hot water will have percolated and circulated, highly altering the solidified material, dissolving and depositing minerals, sometimes reaching land surfaces as geysers and fumaroles. In other areas, around the margins of the Pacific, equal amounts of old oceanic crust will have been swallowed up, descending below the island arcs and continental margins. There it will have been transformed by high pressures and heated up together with trapped seawater and seafloor sediments. At still greater depth, material which has already been through these processes will have started to melt, and the melts will have begun to rise and will have flowed out on the surface, adding to the already huge piles of volcanic products and continually refuelling the Pacific "ring of fire." At all types of plate boundaries, fracturing, folding, uplift, and subsidence will have continued, together with earthquake activity and concomitant surface effects. This is the third way of looking at plate tectonics: from the point of view of the geological processes at plate boundaries (Fig. 23) and their cumulative effects on the rock complexes involved over long periods of time.

Thus, plate tectonic theory has stimulated a search for geological corollaries of plate boundary activity in the rock record of the continents. This search has revealed exposed fragments of ancient oceanic crust, the truncated remnants of the lower parts of volcanic island arcs, supposed equivalents of sediments scraped off the tops of descending lithospheric slabs, exhumed complexes with mineral assemblages indicating the high pressures but comparatively low temperatures expected in subduction zones, and a host of other features showing complex histories thought to be related to evolving plate tectonic environments. The existence of such corollaries has lent land geology a new excitement of discovery, since the continual reworking of the materials out of which the Earth's crust is made has provided visible evidence of processes which today are taking place deep within the crust and on the largely inaccessible ocean floor.

IV. CRUSTAL EVOLUTION

The rocks making up the Earth's crust are consolidated aggregates of minerals, which are the basic chemical components of the solid Earth and are practically all crystalline, with compositions that are fixed or vary within well-defined limits. Crustal rocks are mainly composed of silicate minerals con-

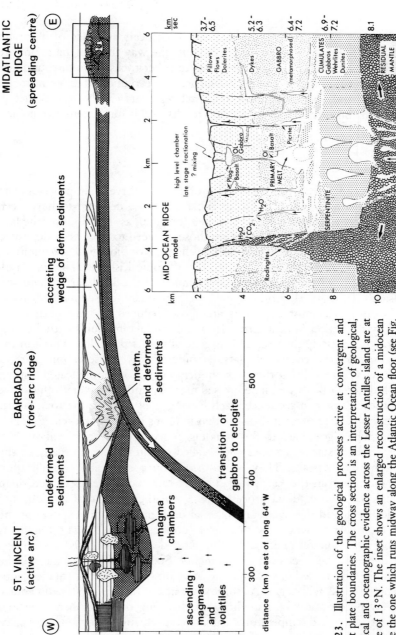

Fig. 23. Illustration of the geological processes active at convergent and divergent plate boundaries. The cross section is an interpretation of geological, geophysical and oceanographic evidence across the Lesser Antilles island arc at a latitude of 13°N. The inset shows an enlarged reconstruction of a midocean ridge like the one which runs midway along the Atlantic Ocean floor (see Fig. 21). (Main cross section from Bott, 1982; inset from Dietrich, in Committee on Marine Geosciences (CMG), 1983, by permission of the authors.)

taining Al, Fe, Mg, Ca, K and Na. Every rock bears evidence of its origin and history in either the minerals it contains, the way the mineral grains are arranged, or the relation of the rock mass to other rocks in its surroundings. By piecing together the fragmentary evidence, the evolution of different rocks, rock bodies, and, eventually, whole crustal segments can be reconstructed. This section considers some of the main results of this fundamental geological activity.

A. The Rock Cycle

To a first approximation, rocks are produced by three main processes and fall into three main groups. Igneous rocks are those deduced to have been formed by crystallization from a rock melt (magma), as, for instance, the lava from a volcano. Sedimentary rocks are those deduced to have been deposited in air or water by mechanical transport or chemical or biological precipitation, as, for instance, the coral reefs around many tropical islands. The unconsolidated materials which commonly cover much of the land surface and sea-floor (sand, mud, gravel, moraine, etc.) are potentially the sedimentary rocks of the future. Metamorphic rocks are those deduced to have been transformed by the action of heat, pressure and/or circulating fluids without passing through a melting process. This is generally caused by deep burial in the Earth's crust in the course of plate tectonic activity. Observation of most rocks, however, reveal that they have undergone a more complicated history, in which different processes, igneous, sedimentary and/or metamorphic, have followed each other in sequence, each leaving their trace in the rock fabric.

It is customary to present this situation in terms of the rock cycle (Fig. 24), a concept which goes back to the founding of geology as a science about 200 years ago. For instance, igneous rocks called granites make up many of the world's mountain peaks. They were formed by cooling and crystallization from a melt deep in the Earth's crust. Plate tectonic processes caused them to be uplifted, and external processes (e.g., weathering, glacial erosion, and landslides) removed the overlying material until they became exposed at the surface today. At present, they themselves are being eroded away; fragments of granite and mineral grains are being altered and transported to the sea, where they will be deposited as sediments. Soon, these sediments will be buried under more sediments and the long process of transformation into sedimentary rocks will begin. At some time in the future, these sedimentary rocks may become involved in the underthrusting of a continent (see Fig. 23), the pressures and temperatures will increase, chemical reactions take place and minerals recrystallize, transforming them into metamorphic rocks, which in turn may at some stage return to the surface, and so the cycle continues. The completion of an evolutionary sequence like the one just described can take anything between 10 and 100 m.y.

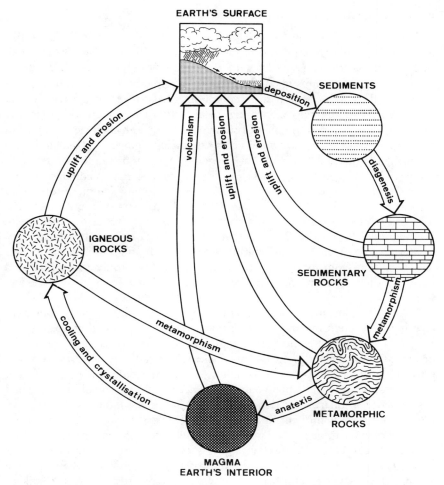

Fig. 24. Schematic representation of the rock cycle.

Although there are only a limited number of cyclic routes, there are almost an unlimited number of ways in which rocks can evolve through a sequence of cycles and each step tends to obliterate the last, so it is rarely possible to go back more than a few steps on the basis of the rock's present state. Similarly, there are almost an infinite number of ways in which large masses of crustal rocks can evolve with time within the plate tectonic framework and we are only just beginning to be able to reconstruct crustal evolution in the far distant past. However, the two main types of crust, oceanic and continental, show quite different developmental patterns and these can serve as a basis for further discussion.

B. Oceanic Crust

Oceanic crust forms at the active mid-ocean ridges, which build an almost continuous system around the globe, about 80,000 km long and standing 2–3 km above the average level of the ocean floor (Fig. 21). The global midocean ridge system is marked by shallow seismicity, active volcanism, and high heat flow. Observation from submarines have revealed numerous fractures parallel to the ridges, forming a deep rift valley in the ridge crest at many localities, marked by piles of submarine lavas with their typical bulbous forms (pillow lavas) and hot water vents supporting strange deep-sea life forms. Seismic and gravity data reveal that anomalous crust and mantle material with no clear Moho occur at depth below the ridges, and the paleomagnetic and stratigraphic data shows that the ocean floor is moving symmetrically away from the ridge crests. These are some of the phenomena that have led to the idea of oceanic crust forming mainly by igneous processes at divergent plate boundaries. Magma rises from the upper mantle along fractures formed by the moving apart of the plates, fills the spaces, cools and solidifies at different levels, or pours out as lava on reaching the sea floor, only to be fractured and torn apart in turn and intruded by more magma from below (see Fig. 23).

Away from the spreading ridges, the oceanic crust shows a very consistent layering of different materials with different seismic velocities (Fig. 25a). Layer 1 ($v_P \sim 1$–3 km/sec) is thought to represent the seafloor sediments: it is thin or absent along the ridge crests and thickens outwards. Layer 2 (v_P 3.4–6.4 km/sec) is supposed to consist of pillow lavas, lava–sediment mixtures, and lava-filled fractures and small chambers. These rocks are mainly fine-grained basalts, poor in quartz and rich in Fe- and Mg-bearing minerals. Layer 3 (v_P 6.4–7.2 km/sec) is thought to consist of deep-seated equivalents of the lavas which cooled more slowly in larger magma chambers and, after solidification, were metamorphosed by the continuing high temperatures and fluid migration. These rocks are coarse-grained gabbros and amphibolites, often showing compositional layering and becoming denser downwards, due to the sinking and accumulation of heavy Fe- and Mg-rich minerals such as olivine. The olivine-rich rocks at the base of the crust grade into the olivine-rich rocks of the mantle, which remained after the magmas were generated and migrated upwards, a transition (the Moho) which is generally ~ 8 km below the ocean floor.

Some of the above ideas have been confirmed by the Deep-Sea Drilling Project, which has now drilled and recovered cores from many hundreds of holes in the ocean floor, mainly penetrating layer 1 but, in some places, also the upper part of layer 2. Dating the layer 1 sediments in the cores, using the enclosed remains of microorganisms, allows the age of the oceanic crust over large parts of the Earth to be estimated. In the Atlantic, for instance,

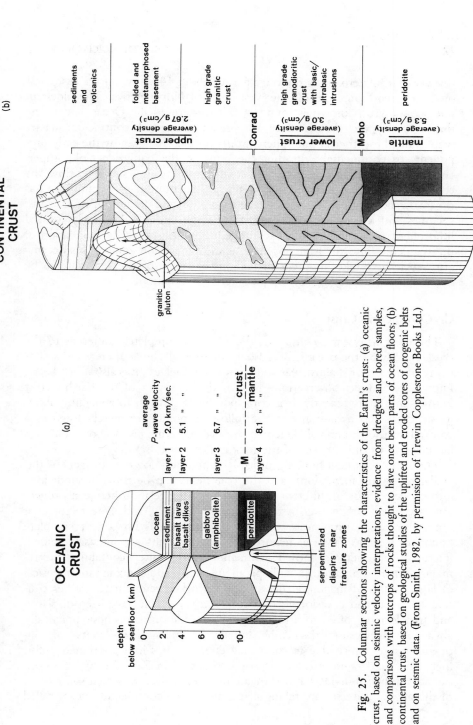

Fig. 25. Columnar sections showing the characteristics of the Earth's crust: (a) oceanic crust, based on seismic velocity interpretations, evidence from dredged and bored samples, and comparisons with outcrops of rocks thought to have once been parts of ocean floors; (b) continental crust, based on geological studies of the uplifted and eroded cores of orogenic belts and on seismic data. (From Smith, 1982, by permission of Trewin Copplestone Books Ltd.)

this varies from zero along the ridge axis to about 140 m.y. off the east coast of the United States and off the west coast of North Africa. The oldest *in situ* oceanic crust known is found in the western Pacific Ocean and is less than 250 m.y. old. At that time, the present, dispersed, continental masses were welded together in one supercontinent, called Pangaea by the early proponents of continental drift, but the crust of the equivalent, old, super-circum-Pacific subduction zones, by underthrusting and sinking below the continental margins and volcanic island arcs. Evolution of oceanic crust is thus a cyclic process: it rises as magma at the mid-ocean ridges; it cools, solidifies and moves passively outwards on the back of the newly formed lithospheric plate; and it sinks back, together with its attached foundation of upper mantle, into the asthenosphere at the subduction zones. This oceanic crust cycle may take hundreds of millions of years to complete but can be much shorter, depending on the plate tectonic situation.

C. Continental Crust

The crust below the continents (Fig. 25b) is completely different to the crust underlying the oceans. Because it underlies all the land areas (as well as many of the major shallow seas and continental shelves), it is also much better known from direct observation—in fact, the only part of the Earth's crust known until quite recently. It is correspondingly difficult to generalize about its structure and evolution and the following points are merely listed as a contrast to the thumbnail sketch of the oceanic crust given above and as a background to the concept of the rock cycles.

The main continent-building process is the one known as orogenesis (literally, mountain formation), and the continents can be roughly subdivided into orogenic belts of different ages. Although the relation between orogenesis, a concept originating in the nineteenth century, and plate tectonics is not always clear, orogenesis seems to be associated with convergent plate boundary systems. Typical features of orogenic belts are intensely folded, fractured, and metamorphosed sedimentary rocks (often containing the remains of marine organisms); widespread volcanism and concomitant intrusion of large bodies of magma at depth; and reworked remnants of earlier continental crust (and occasionally of oceanic crust) which have been displaced and transformed, often to a high degree. In addition, orogenesis practically always involves crustal thickening, initially producing deep "roots" of continental crust, which due to buoyancy effects causes long-continued uplift, high relief, and rapid erosion. At any one period in Earth history, the continents can be subdivided into elongate orogenic zones, in which some or all of the above processes are taking place, and stable continental masses called cratons, often covered by shallow seas or low-lying land and with little volcanism. The present continental crust can be subdivided into zones in which orogenesis has taken place during the last 100 m.y., continuing in places up to the present time and including most of the world's high mountain chains, and cratonic areas, which have remained relatively stable throughout this period (Fig. 26).

Under the thin discontinuous veneer of young sediments, the craton can itself be further subdivided into orogenic zones and stable blocks during earlier periods, back in time until about 2500 m.y. ago. Before this, the subdivision into mobile and stable zones breaks down, and the oldest cores of the cratons show mainly rocks formed by igneous and metamorphic processes taking place continuously or in sequence over wide areas and long periods of time. This illustrates one of the main differences to oceanic crust: continental crust is not being continually generated and destroyed; it was mainly formed by enigmatic processes at a very early stage in Earth history and has evolved since then by a reworking of the already formed materials. For the last 2500 m.y., only relatively small amounts of new material, in the form of magma or subducted magmatic arcs, have been added to the continental crust; the other materials involved were already present as sedimentary, igneous, and metamorphic rocks from earlier orogenic events, which themselves involved mainly reworked remnants of the primordial crust. The evolution of continental crust can thus be conceptualized in terms of orogenic cycles, each lasting tens to hundreds of millions of years.

The Moho below the continents is very well defined but lies deeper and is more variable than the base of the oceanic crust. It is deepest below the youngest orogenic belts (the present-day mountain chains of the world), reaching 60 km below sea level in places, whereas the normal thickness under the cratons is ∼ 35 km. In some areas of the continent, where rifting similar to that at a mid-ocean ridge is taking place, the stretching has reduced the thickness to 10 km (as also at old rifted margins, e.g., on both sides of the Atlantic). In spite of the complexities of orogenesis, the continental crust does show a rough layering which can be read from seismic and gravity data (Fig. 25). In the cratonic areas, the upper layer consists of flat-lying sediments, sedimentary rocks, and/or volcanics (lavas, ash layers, etc.) which are collectively called the cover. Over large areas, however, this is absent and the underlying basement comes to the surface: metamorphic and igneous rocks of earlier orogenic episodes with an average composition of granite (rich in quartz and feldspar and poor in Fe and Mg minerals, with a mean density of 2.7 g/cm^3 and $v_P = 5.9-6.5$ km/sec). This represents the exposed part of the upper continental crust, whose base is marked by a weak seismic reflector (Conrad discontinuity) at 10–20 km depth. Below it, we have the lower crust,

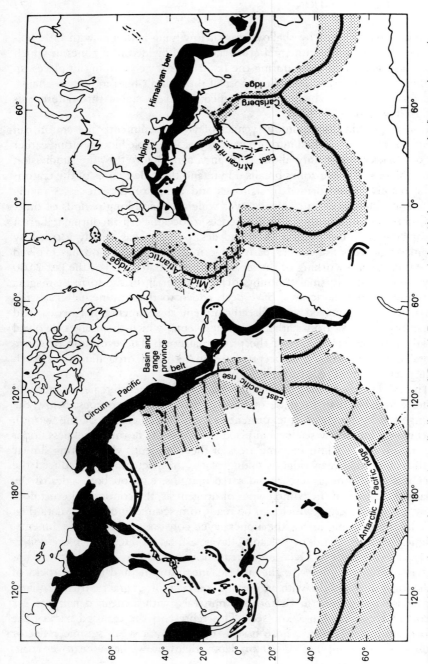

Fig. 26 World map showing the approximate distribution of Cenozoic-Recent orogenic belts (dark ornament on land), cratonic areas (unornamented land) and ocean floor (light ornament on both sides of the midocean ridges). (From Bott, 1982, by permission of the author.)

of higher density (~ 3.0 g/cm^3) and higher P-wave velocity (6.5–7.6 km/sec), thought to be made up of dry, high-pressure equivalents of upper crustal rocks and igneous material intruded from the upper mantle. At the base of the lower crust (Moho), pressures of 10 kbar and temperatures of 600–700°C are expected, and the rocks are thought to be mainly metamorphic, belonging to a large group known as granulites.

In summary, we can say that crustal evolution in general, and rock evolution in particular, is best thought of in terms of geological cycles related to global plate tectonics and taking place over time periods measured in tens or hundreds of millions of years. Materials within these cycles, however, usually go through stages of much more rapid change and long periods of comparative calm. The stages of rapid change occur whenever the materials are involved in the boundaries between the lithospheric plates, whether divergent or convergent. The periods of calm coincide with incorporation within the plates themselves, or with the return to the deep mantle. Thinking of the future, and of the artificial materials, such as radwaste, which are to be introduced into the crust by Man, it is clear that the aim must be to enter the geological cycles at points which guarantee the longest possible period of calm. This means, on the global scale, that mid-plate situations are to be preferred, away from ridges, rifts, trenches, island arcs, and volcanic chains and outside the orogenic zones of the last 100 m.y.

V. THE WATER CYCLE

Although much of this chapter has dealt with the interior of the Earth, at several points we have indirectly singled out the Earth's surface as of special interest. The Earth's surface is a complex interface between rock, water, and air and is continually being sculptured and remodelled by internal forces directing the cycles of tectonic and volcanic events and by external forces directing the cycles of weather and climate. It was at this interface that life originated and evolved, creating the biosphere as we know it today (see Chapter 4), and it is at this interface that Man lives and works and creates environmental problems like the one which is the focus of this book. In fact, the following chapters are almost exclusively concerned with, geologically speaking, transient processes and interactions at or within a few hundred metres of the surface of the solid Earth. The role of water in this surface or near surface system is of critical importance for the understanding of many geological processes, including all types of plate boundary activity. Here, we introduce it conceptually as another, albeit much more rapid, cyclical motion, the global water cycle.

On a global scale, the Earth's surface may be viewed as a large distillation unit run by solar energy, in which water is continually evaporated, mainly from the oceans, transported in the atmosphere, and condensed and precipitated. There is a net loss from the oceans and a net gain on land which is compensated by water flow from land to sea, most obviously in the form of rivers (Fig. 27).

The runoff from land to sea under the influence of gravity is also the main agent of sediment transport; rocks exposed at the Earth's surface are weathered and biologically disintegrated, forming in most areas an unstable, superficial layer of unconsolidated material, which moves downslope by various processes until it enters the rivers and streams and is washed or dissolved away. This complex of processes leading to the denudation of the land and the deposition of sediment in the oceans is treated in more detail later (see Chapters 6 and 7). The time taken for water molecules to complete such a cycle can be only months or years. However, water can get trapped for thousands of years or more at certain points in the cycle. There are three main traps: the ocean itself (particularly the deep ocean), the ice caps and glaciers, and the pore spaces in the rocks of the upper few kilometres of the Earth's crust. In these traps, water movement can be so slow that it can hardly be measured, and it is precisely these environments which are under consideration for radwaste disposal.

A. The Ocean Water Trap

The largest water trap is the ocean itself, with a total volume of 1350×10^6 km^3. Although the shape and size of the ocean basins has changed considerably over the last million years, there is good evidence that the volume and composition of the sea water has not. Sea water composition is measured in terms of salinity, which is the total content of dissolved constituents (average value 3.45%), and there is an intimate relation between salinity, temperature, and density which determines the large-scale behaviour of the water masses (see Chapter 12). At depths of 0–200 m in the open ocean, these are strongly influenced by weather and climate. In this surface layer of wind-mixed waters, major circular currents dominate water movement in both the Atlantic and Pacific oceans (gyres), with maximum velocities of 15 m/sec in places. At a depth of 200–600 m comes the thermocline, a stable layer which prevents intermixing of surface and deep ocean waters and which is marked by a rapid drop in temperature to that of the deep layer ($< 5°C$) and a concomitant change in salinity and increase in density. At depths > 600 m, sluggish movement of large water masses of relatively constant temperature and salinity is observed. Downsinking through the thermocline occurs in Arctic and Antarctic regions, and upwelling occurs along the west coasts of the continents, where surface currents flow away from the coastline. Dividing the total ocean

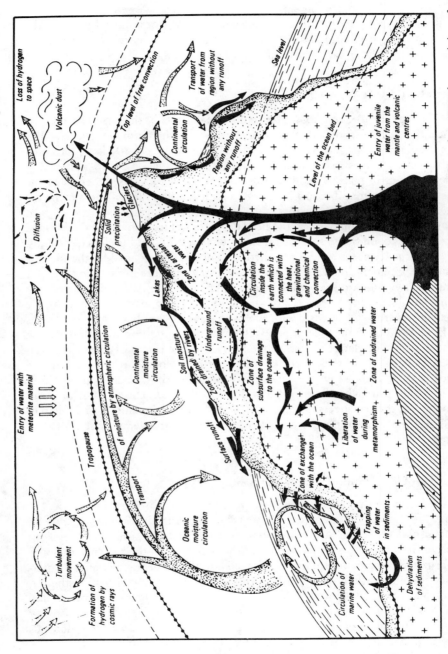

Fig. 27. Imaginative representation of the general circulation patterns of water in nature. (From Pinnecker, 1983, by permission of Cambridge University Press.)

water volume by the volume of water evaporated from and returned to the ocean every year (40,000 km^3), gives an average "residence time" for water molecules in the ocean trap of about 40,000 yr.

B. The Glacial Water Trap

The second large trap for water molecules consists of the world's ice sheets (Antarctica and Greenland) and, to a lesser extent, the mountain glaciers. The total volume of frozen water is at present about 3.0×10^7 km^3, although during the ice ages it was much greater (estimated at 7.5×10^7 km^3). Residence times in glaciers are measured in hundreds of years, whereas in the ice sheets they may be 100,000 years or more. The rate of movement depends on many external factors, such as bedrock topography, terrestrial heat flow, and climate, but also on the flow properties of ice, which are determined by internal conditions, such as pressure, temperature, and presence of fluids. In the large ice sheets, the internal conditions are best defined by the melting point of ice at the relevant pressures; "warm" ice, near the melting temperature, flows much more rapidly than "cold" ice. The Antarctic and Greenland ice sheets, for instance, are generally cold, but in many areas the ice becomes warm and coexists with meltwater at the very base, where it forms a kind of lubricating layer. It is possible that this condition may underly the phenomenon of glacier surges, sudden periodic movement of large valley glaciers, particularly in the Arctic, in which surface velocities of 5 km/year and more are attained. However, the surface velocities of normal "warm" valley glaciers are generally of the order of 10–100 m/year. The amount of water trapped in the ice sheets depends on global climatic conditions (see Chapter 13) and is perhaps the most variable component of the water cycle on a long time scale.

C. The Underground Water Trap

The amount of water trapped underground is estimated at about 8×10^6 km^3. This occurs in two environments, in the thin discontinuous layer of unconsolidated material which covers most of the land surface (the regolith) and in the underlying rock complexes (the bedrock). In the regolith, the rock and mineral fragments are loosely packed and the percentage of air or water-filled volume is often 20–50% (porosity). This is not always part of the underground trap, since flow velocities of up to 15 m/day are common and the residence time can be less than a year. In the underlying bedrock, porosity is generally lower and the ability of a rock to allow water flow often depends more on the degree of cracking and fracturing. This property of large rock units is known as permeability: some rock units are permeable because of high porosity (sandstones) but others, sometimes with porosities of 20–50% (e.g., shales), are impermeable because the pores are small and not interconnecting and because they do not fracture easily. Other rock units of low porosity such as limestone or granite (porosity 1–2%) are highly permeable because their brittle nature allows widespread cracking. In the most permeable rocks (acquifers), rates of water flow at depth may be up to 1 m/day but are generally around a few centimetres per day; in impermeable layers (aquitards), water movement is extremely slow, generally < 1 m/year.

In general, permeability decreases with increasing depth. Waters have been found in the deepest of oil wells (9 km depth), where they are usually strongly mineralized and quite hot, with maximum flow speeds of < 1 cm/year. However, even this is not a negligible rate on the geological time scale, and geologists consider water as an important component within the whole of the crust and as playing an important role in almost all geological processes.

VI. CONCLUDING REMARKS

Geologists consider the Earth's crust to be a constantly changing skin on a dynamic body, which evolves on a time scale of millions of years. The skin forms the outer layer of a series of stiff plates which are in continual motion, driven by the internal energy of the Earth. Where the plates are torn apart, move past each other, or collide, mobile zones develop which, even on the shorter time scale relevant to radwaste disposal, show increased activity, including large earthquakes, active volcanoes, geothermal fields, rapid uplift or subsidence and high relief. In all these processes, water plays a role which can hardly be overestimated. An understanding of crustal evolution and hydrologic processes is thus an essential basis for developing disposal concepts, particularly for radwastes requiring very long isolation. The aim of this chapter was to provide a brief survey and an entry into the geological literature.

SELECTED LITERATURE

Internal structure of the Earth: Elder, 1976; McElhinny, 1979 (pp. 1–58); Bolt, 1982, Bott, 1982; Press and Siever, 1982 (pp. 319–330, 393–418); Smith, 1982 (pp. 36–49, 141–161).

Global plate tectonics and crustal evolution: Dickinson, 1972; Le Pichon *et al.*, 1973; Marvin, 1973; Smith and Briden, 1977; Tarling, 1978; McElhinny, 1979 (pp. 567–593); Burchfiel *et al.*, 1980; Bott, 1982 (pp. 130–139); Gass, 1982; Press and Siever, 1982 (pp. 165–177, 219–273); Westbrook, 1982; Owen, 1983; Siever *et al.*, 1983.

Water cycle and fluids in the Earth's crust: Sugden and John, 1976; Turekian, 1976; Fyfe *et al.*, 1978; Freeze and Cherry, 1979; Smith, 1981 (pp. 291–296, 311–324); Press and Siever, 1982 (pp. 129–154); Pinnecker, 1983.

CHAPTER 4

Geological Time

I. INTRODUCTION

At several points in the last chapter, we mentioned spans of time which are almost inconceivable. When we consider the span of human history, say the last 5000 years, we feel very far removed from the founders of agriculture and civilization in ancient Egypt and Mesopotamia. It takes an enormous effort to think back to *Australopithecus*, our early forebears in central Africa, some 1 million years ago, a span 200 times as long. To project 1 million years into the future seems an almost impossible task, although that is what this book is about! When it comes to plate tectonic processes and crustal evolution, and to the origin of the Earth itself, the time required seems to be infinitely long, just as the stars and galaxies are for all practical purposes infinitely far away. Yet it is not infinite. It is, in fact, finite and measurable, and some concept of the geological time scale, the history of the Earth and the evolution of life is necessary to keep human problems, such as radwaste disposal, in a proper perspective. Here we discuss briefly how such long periods of time can be measured and how the history of life has been reconstructed.

II. DATING METHODS

The biblical estimate of the creation of the world some 4000 years BC kept geology in a straitjacket until well into the nineteenth century. However, from the mid-eighteenth century onwards, lone voices were raised that this time was too short. Around 1830, the scientific rationalization known as actualism, the theory that processes observed today operated in the same way in the past and will continue doing so in the future, was introduced and extended to geology by Charles Lyell, the "father of geology" and close friend and advisor of Charles Darwin. This led the way to various estimates of the age of the Earth based on, for instance, the rate of cooling of a hot sphere the size of the Earth, the rate of deposition of sediments in deltas and in glacial lakes, the rate of evolution of animal and plant species found in sedimentary rocks, or the amount of salt in the sea. Although primitive, all these attempts showed that Earth history must be measured in at least tens of millions, not in thousands of years. At the same time, the synthesis of geology and biology which led to Darwin's theory of evolution provided a tool for the relative dating of sedimentary rocks, using the enclosed fossils (remains of plants and animals) to compare material of unknown age with the painstakingly constructed and ever more refined evolutionary scheme (biostratigraphy).

The big breakthrough, however, came at the turn of the century with the discovery of radioactivity. It soon became clear that the time at which minerals containing radioactive elements, such as uranium, crystallized could be measured if the decay constant and the amount of the stable daughter isotope could be determined (radiometric age dating). First measurements on uranium ores gave ages of the order of 500 m.y., but before World War II, the measurement techniques were not capable of giving more than rough estimates. Since then, the techniques have become steadily more refined and an accurate and reproduceable geological time scale has been constructed and fitted to the earlier scheme based on biological evolution.

Although biological evolution and radioactive decay give a quantitative framework, the history of the Earth is based on a third, more qualitative source of information, the field relations. These describe how the different rock units are arranged relative to each other on many different scales, from global plates down to aggregates of minerals under the microscope. The geologist functions like a detective searching for clues and reconstructing the most likely sequence of events. Some clues are obvious and unambiguous, others are indirect and based on delicate observations and fine distinctions, often ambiguous. A compilation of the different methods and arguments used would cover the branch of earth science known as historical geology, whose aim is to elucidate as completely as possible the history of our planet. In the following, we look at the three main dating methods in more detail, before reviewing the main results.

A. Biostratigraphy

Biostratigraphy can be defined as the use of the fossils enclosed in sedimentary rocks to characterize successive intervals of geological time. The fact that successive strata contain different assemblages of fossils was discovered by William Smith in England and Georges Cuvier and Alexandre Brongiart in France at the beginning of the nineteenth century. This enabled these workers to construct the first geological maps in areas of poor exposure, since widely separated rock outcrops could be identified as belonging to the same sedimentary formation (correlated) even when the rocks themselves did not show identical characteristics. The use of fossils as a means of correlation became widespread long before Darwin's *Origin of Species* provided a theoretical justification and allowed the technique to be applied also to regions of complicated structure, such as the world's orogenic belts. Later, biostratigraphy became one of the main tools in the exploration for oil and gas, permitting the demonstration of correlations between different boreholes on the basis of a detailed study of the microfossils in the drilled formations. The most useful technique involves the definition of concurrent-range biozones (Fig. 28), but paleontological events (e.g., first or last appearance of a partic-

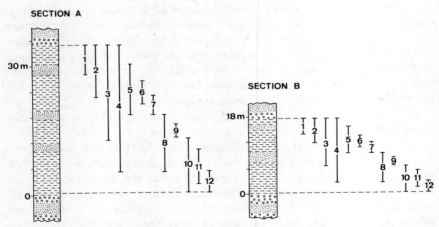

Fig. 28. One method of biostratigraphic correlation of sedimentary successions. The ranges of fossil species 1–12 in two borehole logs show that they represent the same time period although the sedimentation rate in section B was much less (concurrent zones). (Data from Eicher, 1968.)

ular species) and many types of statistical analysis are at present in use. The basic principle is that a given assemblage of fossils is typical of a given small segment of Earth history, i.e., a given "instant" of geological time. Two main problems make biostratigraphy only useful as a dating method in conjunction with other techniques. First, it is a purely relative dating method, indicating whether one rock sequence is older than, younger than, or the same age as another. Second, the fossil assemblage is often strongly affected by the environmental conditions (e.g., open ocean, nearshore, alluvial plain) (see Chapter 7), so that lateral correlation also has to take into account the ancient geography and climatic zones. Hence, biostratigraphy can be more successfully applied in marine environments using free-swimming or floating organisms and is extremely difficult in nearshore and continental environments using organisms adapted to specific ecological niches.

B. Radiometric Age Dating

The first problem of biostratigraphy, its relativity, is usually approached by attempting to put an absolute age on some of the rocks in a given sequence and by developing methods of extrapolating between the dated points. Within this field, the techniques of radiometric age dating take up a central position, although they are more easily applied to igneous and metamorphic rocks. Radiometric age dating is based on the decay rates of naturally occurring radioisotopes. The ones generally used for this purpose are ^{238}U, ^{235}U, ^{87}Rb, ^{40}K, ^{230}Th, and ^{14}C, but many others may have special applications as measurement techniques (e.g., mass spectrometry, fission track dating, and neutron activation analysis) improve. The basic physical law underlying the method is that the number of atoms disintegrating per unit time ($-dN/dt$) is proportional to the total number of radioactive atoms (N) which are present, giving the exponential decay curves for the parent (radioactive) isotope and the corresponding accumulation curve for the daughter (radiogenic) isotope (Fig. 29). Knowing the decay constant (see Chapter 1, Section III,A) and the ratio of parent to daughter isotope allows the time of formation of the parent to be calculated, if two basic conditions have been fulfilled: (1) the system has remained closed for both the parent and the daughter isotope since the mineral in which they occur formed and (2) the daughter isotope was not present in the mineral at the time of formation. Since these conditions are rarely fulfilled in nature, radiometric age dating is often based on reasonable assumptions about the original conditions or on deductions of the evolution of the rock mass made on the basis of other information. If condition (1) is not fulfilled, only the case in which the system is open with respect to the daughter isotope can be used (e.g., concordia diagram) (see Fig. 29b), but the isotopic ratios in any case give clues as to the amount and rate of radionuclide migration

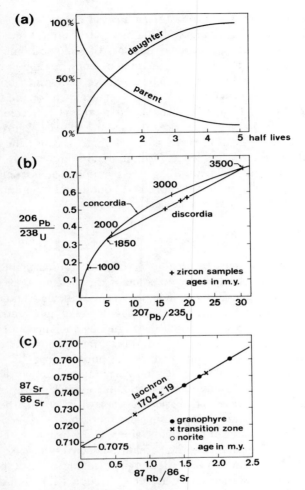

Fig. 29. Aspects of absolute dating using the decay of natural radioactive isotopes (radiometric age dating): (a) decay curve of a parent radioisotope and the corresponding accumulation curve of the daughter nuclide; (b) the concordia curve for uranium–lead isotopic compositions in a closed system and the plot of discordant zircons from southern Minnesota, indicating that their history included an episode of lead loss (about 1850 m.y. ago); (c) whole-rock isochron for comagmatic igneous rocks from Sudbury, Ontario, containing different quantities of rubidium/strontium isotopes with the same isotopic ratios. The data points define a line whose slope indicates that the rocks crystallized 1704 ± 19 m.y. ago. (Data from various sources.)

in the crust (see Chapter 11). If condition (2) is not fulfilled, the original amount of the daughter present and the age of formation can often be determined simultaneously using concentration variations in the original parent isotope (e.g., isochron diagram) (see Fig. 29).

C. Field Relations and Other Methods

Biostratigraphy and radiometric age dating are well-defined disciplines with appropriately developed techniques and philosophies. However, as pointed out above, they often face basic problems which can only be approached by collecting evidence from the whole geological picture. This can be best illustrated with reference to the field relations in a particular area (Fig. 30), in which the two most general "laws" are clearly illustrated: the law of superposition (in an undisturbed sequence of sedimentary rocks each layer is younger than the one below and older than the one above it) and the law of cross-cutting relationships (in a layered rock complex which is discordantly cut through by a layer of another type, the latter is the youngest). However, the field geologist uses a whole spectrum of arguments in building up the geological history of an area, and any kind of data on any scale of observation

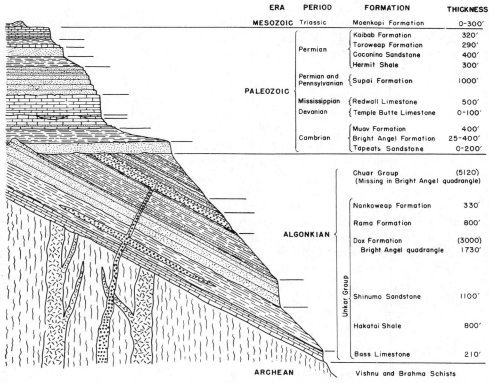

ERA	PERIOD	FORMATION	THICKNESS
MESOZOIC	Triassic	Moenkopi Formation	0-300'
PALEOZOIC	Permian	Kaibab Formation	320'
		Toroweap Formation	290'
		Coconino Sandstone	400'
		Hermit Shale	300'
	Permian and Pennsylvanian	Supai Formation	1000'
	Mississippian	Redwall Limestone	500'
	Devonian	Temple Butte Limestone	0-100'
	Cambrian	Muav Formation	400'
		Bright Angel Formation	25-400'
		Tapeats Sandstone	0-200'
ALGONKIAN		Chuar Group (Missing in Bright Angel quadrangle)	(5120')
		Nankoweap Formation	330'
		Rama Formation	800'
		Dox Formation Bright Angel quadrangle	(3000) 1730'
	Unkar Group	Shinumo Sandstone	1100'
		Hakatai Shale	800'
		Bass Limestone	210'
ARCHEAN		Vishnu and Brahma Schists	

Fig. 30. Diagrammatic summary of field relations in Grand Canyon, Arizona, showing the succession from oldest (lowest) to youngest (highest) rock units, with occasional unconformities indicating periods of tectonism followed by uplift, erosion, and renewed subsidence and sedimentation, and discordant rock bodies indicating periods of igneous intrusion.

may be taken into account. Techniques which are most commonly used in a historical context include sedimentary facies analysis (reconstruction of ancient sedimentary basins), magnetostratigraphy (magnetic polarity changes in sedimentary rocks, lavas and deep sea cores), mineral texture and rock fabric analysis (microscopic–submicroscopic study of mineral growth and alteration sequences), time-span analysis using a whole spectrum of techniques including sedimentation rates, rates of organic evolution, isotope ratios as climatic indicators, and volcanic ash chronology, and any other kinds of information indicating relative times of formation. It is this combined accumulation and interpretation of all types of earth science data which has resulted in the present understanding of earth history and which will result in a continual modification and refinement of historical geology in the future.

III. EARTH HISTORY

Study of isotope systematics on terrestrial and lunar rocks and on meteorites shows that the Earth originated 4600 m.y. ago, probably at the same time as the other inner planets, by the condensation of an interstellar cloud and the gradual accretion of small proto-planets into the present bodies. Energy from impacts, from gravitational collapse, from radioactive decay and from the formation of the dense core contributed to raise the temperature above the melting point, but cooling soon produced a primitive crust, which continued to be intensely bombarded by condensed fragments. Insights into this early history of the Earth are best obtained from the other planets, and from the Moon, where the youngest rocks (the mare lavas) are between 3200 and 3800 m.y. old and where the main epoch of cratering ended 4000 m.y. ago. On Earth, the oldest rocks are 3800 m.y. old and signs of Earth's primordial development have been erased by the subsequent dynamic evolution of the crust, which did not take place on the other planets and on the moon. Throughout Archaean time (up to about 2500 m.y. ago) (Fig. 31), the crust developed by processes akin to plate tectonics but probably more rapid and fluid, dominated by convection currents, thermal jets and widespread volcanism. The end result was the accretion of light material with an average composition similar to granite, eventually forming the continents as large buoyant masses at the surface.

A. History of Life

Some of the oldest known rocks on the Earth are sedimentary, so there is no doubt that it had cooled sufficiently for water to exist on its surface by 3800 m.y. ago. At this time, there was no free oxygen in the atmosphere and the composition of seawater was quite different from its composition today. The first chemical reactions and the building of the organic compounds necessary for the development of living organisms were taking place under the influence of intense ultraviolet radiation. The remains of early microorganisms appear first in rocks about 3500 m.y. old, and algal colonies (stromatolites) became quite common after about 2500 m.y. ago, which marks the beginning of the Proterozoic (Fig. 31). Photosynthesis as a means of maintaining life processes seems to have developed early in the Proterozoic and resulted in a drastic change in the atmosphere, with steadily increasing amounts of free oxygen from 2000 m.y. ago onwards. Marine life flourished and diversified, and organisms increased in complexity and, towards the end of the Proterozoic, developed into a spectrum of species related to present-day jellyfish, sea pens, worms and possibly sea urchins. Since this time, ~ 1000 m.y. ago, the composition of seawater has changed little and organisms with calcium carbonate shells and skeletons become increasingly common.

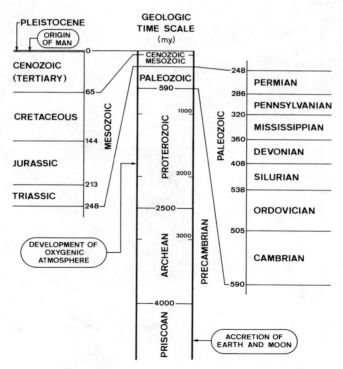

Fig. 31. A skeleton geologic time scale showing the main subdivisions of Phanerozoic time (from 590 m.y. onward).

The end of the Proterozoic and the beginning of the Phanerozoic (at 570 m.y. ago) saw a great blossoming in the variety of marine life forms, as well as the sudden development of species with hard parts in practically every group of organism. In contrast, the land areas remained barren, subject to rapid erosion and desertlike under all climatic conditions. Their greening by the earliest land plants took place about 400 m.y. ago, possibly facilitated by the buildup of oxygen levels in the atmosphere and the creation of an ozone layer, which shielded against the harmful ultraviolet rays (Fig. 32). With the

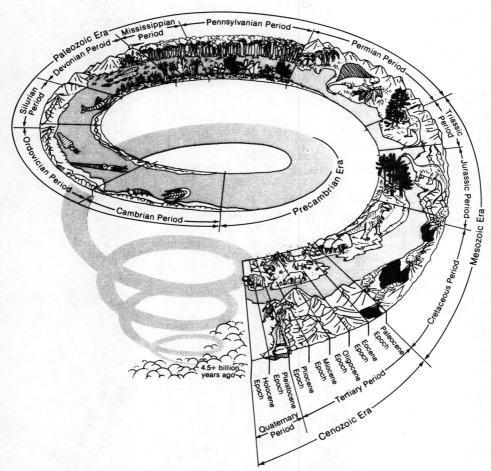

Fig. 32. The Phanerozoic time scale and the evolution of life. (From Press and Siever, 1983, by permission of W. H. Freeman and Co.)

plants, the animals could now invade the new ecological niches on land, first the arthropods (including millipedes and spiders) and then, about 350 m.y. ago, the four-footed vertebrates (amphibians, reptiles). However, this triumph was short-lived by geological standards, and the end of the first epoch of the Phanerozoic (Paleozoic), between 300 and 200 m.y. ago, is marked by the extinction of many of the older species, particularly of reptiles and shallow-water marine organisms. This long-drawn-out process was probably related to extreme climatic conditions (widespread glaciation) and unusual crustal conditions (single continent, shelf areas reduced to a minimum). The result was an explosion of new forms when conditions became more congenial and more stable at the beginning of the Mesozoic, 200 m.y. ago.

The Mesozoic is the age of the dinosaurs, which, together with insects and flowering plants, spread across the huge areas of dry lowland (see Fig. 32). The success of the dinosaurs probably restricted the development of the small ancestral mammals which had also appeared on the scene. In the seas, a new group of organisms, the ammonites, blossomed up, and octopuslike forms became abundant, in addition to many species related to Paleozoic groups. The tremendous profusion of Mesozoic life came to an abrupt end about 65 m.y. ago, in an event which is as enigmatic as it was catastrophic. The groups to suffer most were exactly those which had dominated: the dinosaurs on land, the plankton and ammonites in the sea. All the changes and extinctions seem to have taken place within a time span of 50,000 years, a length of time which makes this unusual event of some relevance to the radwaste problem. There are now various lines of evidence that the impact of an unusually large asteroid, 10 km in diameter, led to such strong environmental changes (ozone depletion, chemical pollution of the ocean, and alteration of atmospheric composition and dust content) that certain groups of species were unable to adapt. The mammals were not one of these—they could not only adapt, they underwent enormous expansion as they moved into every ecological niche left vacant by the doomed dinosaurs. From 65 m.y. ago onwards we enter the age of the mammals, the Cenozoic, and the establishment of the present biological spectrum on land, in the sea, and in the air (Fig. 32).

B. The Last Million Years

Towards the end of the Cenozoic, around 10 m.y. ago, apelike mammals left the forests and migrated out into the grasslands, developing biped loco-motion and using the freed forearms and hands for gathering food, making tools and building shelters. These early hominids (*Australopithecus, Homo erectus*) were widely distributed by the beginning of the Pleistocene, 2–3 m.y. ago, particularly in the warmer areas of Africa and southern Eurasia. The Pleistocene was characterized by climatic variations on a scale that the Earth had not seen for at least 250 m.y., causing the rapid growth and decline of continental ice sheets (see Chapter 13) and the consequent continual displacement of climatic zones. Far from being detrimental, these unusual environmental fluctuations seem to have encouraged and accelerated mammal evolution, so that 1 m.y. ago, the ice-free land areas supported large populations of wild game which the early hunters could catch with the most primitive tools and weapons (Fig. 33). Members of the genus *Homo* spent 99% of this whole period, the Paleolithic, as hunters and gatherers, slowly learning to produce more sophisticated and finely worked tools and weapons out of various rock materials and slowly learning the uses of fire and clothing. By the onset of the last great Ice Age, 120,000 years ago, early *Homo sapiens* (e.g., Neanderthals) could thrive in Arctic climates and tap the huge game reserves of the periglacial tundra. The late Pleistocene saw the rapid devel-

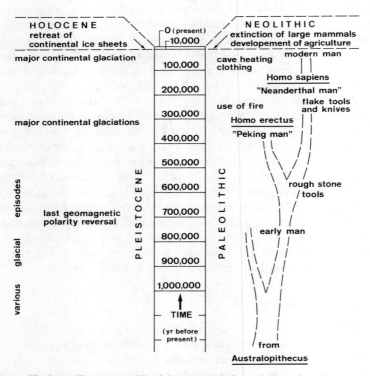

Fig. 33. The last million years of Earth history, with the evolution of modern man (during the archaeological time segment Paleolithic) and the recurrence of continental glaciations (during the geological time segment Pleistocene) (see also Fig. 81). (Data from McAlester, 1968, and Barraclough, 1978.)

opment of many new skills, including sewn clothing, heated rock shelters, finely chipped knives, and cave painting and carving, concomitant with increasing brain capacity considered to be "modern" after 40,000 BP (i.e., with a mean volume of 1500 cc).

The end of the Paleolithic about 10,000 years ago (Fig. 33), coinciding as it does with the end of the last Ice Age and the development of the warm climates of today, was marked by the rapid extinction of most of the large land mammals (perhaps due to over-hunting) and the origins of characteristic patterns of our present civilization: cultivation of plants, domestication of animals, and the introduction of new techniques (grinding, polishing, pottery-making). The development of agriculture, permitting permanent settlements and encouraging the division of labour, between 10,000 and 5000 BP, together with the first use of metal tools and the development of writing (the Neolithic Revolution) is probably the most significant series of events in all of human history. Five thousand years have passed since then, a minute fraction of Pleistocene history, itself an even more minute fraction of the history of the Earth.

IV. RATES OF CHANGE

The main theme of Chapters 3 and 4 has been the geologist's view of our planet as a dynamic body which has evolved continuously over unimaginably long periods of time and which will continue evolving in the future. Study of the properties of the Earth's interior and the rocks of the Earth's crust, together with observation of the processes presently active at the Earth's surface, leads to an ever deeper appreciation of this evolution, and, theoretically, to an ever-improving ability to predict future developments. This is an important aspect of the earth forecasting required for the very long-term planning needed for activities such as radwaste disposal. The question is, shall we ever know enough? Here, we summarize the rates at which the various geological processes take place, as deduced from the rock record. The problem of using this type of information to predict future changes will be reconsidered in Part III.

A. Time-Averaged Rates of Slow Processes

Many geological processes seem to take place at very slow rates which have apparently been maintained for very long periods of time. In such a case, it is usual to determine the rate of change by dividing the net change by the time interval over which this change was observed, and then to convert this to some standard time unit. For radwaste disposal, we are interested in time intervals from 1 yr to 1 m.y., so we shall take as our standard unit 1000 years, a length of time which is short enough to be grasped but long enough to allow important changes to take place by imperceptible processes. As a standard unit of length we shall use 1 m, so that all rates are expressed in metres per millenium (m/1000 years).

The main processes which fall into this category are horizontal plate motion, vertical movement of the Earth's surface (uplift and subsidence), denudation of the land surface (which roughly keeps pace with uplift), and sedimentation on the seafloor and in inland basins. Time-averaging the relative horizontal displacements of parts of the Earth's crust gives mean rates which vary between 1 and 100 m/1000 years, with the lower rates referring to continental fault zones and the higher ones to ocean-floor spreading. Rates of uplift or subsidence have been measured in many parts of the world using repeated accurate surveying or historical, archaeological, and geomorphological data. These yield values of 1–10 m/1000 years in mountainous areas, but local rates of up to 76 m/1000 years have been measured. Erosion rates are generally measured for whole drainage systems by estimating the amount of material carried by the streams and rivers (see Chapter 6, Section III). The average denudation rate of the world's lowlands is ~0.05 m/1000 years, whereas for mountainous areas it is 10 times this figure. However, erosion rates of individual regions can be much higher, up to 13 m/1000 years in some small drainage basins.

Sedimentation rates on the seafloor, that is, the thickness of a sedimentary layer divided by the time over which it was deposited, vary from ~0.01 m/1000 years from deep-sea oozes to 0.1 m/1000 years on the continental shelves, with local values reaching 1 m/1000 years. In contrast, coral reefs build up at rates of 10 m/1000 years and some local inland basins have subsided and filled up with sediment even faster. Surveying these slow continuous processes and considering regional and long-term effects shows that typical rates of geological processes lie between 1 and 10 m/1000 years (or 1 mm–1 cm/year). This applies to other types of movement also, such as the flow of deep groundwater, the rise in sea level with the melting of the Pleistocene ice sheets, the movement of ice particles in the Antarctic ice sheet, and the rise of salt diapirs. Local and short-term rates, however, often far exceed these values, and this leads to the question of the possible significance of violent but rare events.

B. Probabilities of Rare Geological Events

Several types of geological phenomena take place very rapidly—on the geological timescale, instantaneously. Some are sufficiently large-scale to leave a lasting mark on a region of the Earth's surface, either in the landscape or in the sedimentary rocks deposited during or after the event, but they occur only rarely. Three such groups of phenomena have been mentioned earlier: meteorite impacts, large earthquakes and major volcanic eruptions; others include floods, landslides, submarine slides (turbidity, currents) and tsunamis, all of which may be triggered by the first three. For all these phenomena, there seems to be an inverse relationship between the frequency of occurrence and the size of the event, for a given region of the Earth's surface. The probability of large events of this type taking place within a certain length of time is estimated, first, by defining the units of size (e.g., radius of meteorite, volume of landslide, magnitude of earthquake) and, second, by determining a size-frequency relation for smaller events of the same type. The frequency of much larger events is then obtained by extrapolation or statistical analysis (see also Chapter 14).

At any place on the Earth's surface, a meteorite impact of the size which would produce a 1-km-diameter crater is expected every 1400 years, but over longer time periods much larger impacts are likely to occur. For instance, the terminal Cretaceous impact event is expected once every 300 m.y. (probability 10^{-8}/year). Similarly, large earthquakes with an intensity similar to the San Francisco event of 1906 are expected to take place even in cratonic areas at least once every 10,000 years (probability 10^{-4}/year). Small submarine slumps, resulting in a 10-cm-thick turbidite layer on the sea floor, have a frequency of 1–10/1000 years; slumps resulting in a 1-m-thick layer occur every 1 m.y. Very large slumps causing layers several metres thick over large areas of the sea floor and having a significant effect on marine life and water circulation patterns are to be expected every 10 m.y. or so, and, in fact, such beds are regularly found in sedimentary rock sequences. These are just some examples of the geological significance of rare events. In general, it is important to note that there is no sharp dividing line between slow continuous and fast intermittent processes. It is a matter of common observation, for instance, that most erosion takes place during flooding and that the erosive power of many rivers between floods is minimal, so the erosion rates quoted above are long-time averages of the effects of unusual, if not rare, events. It may be that many other apparently slow geological processes in fact take place in quantum jumps (for instance, movement on fractures or rise of salt diapirs) and that continuity and discontinuity are dependent on the time scale being used.

V. CONCLUDING REMARKS

In a scale model of the geological time scale, we can put 1 year equal to 1 sec. An average human life-span would then be approximately 1 min and the beginnings of civilization would be 100 min, or just under 2 hr, ago. The last great Ice Age began yesterday or the day before (33 hr ago), and primitive

Man was first venturing onto the grasslands 1 month ago. The extinction of the dinosaurs, and the beginnings of mammal dominance on land, happened 2.5 years ago but life itself evolved in the ancient oceans of the 1920s and the Earth was born as a planet as Queen Victoria ascended the throne, around 1840. With the same conversion factor, the lithospheric plates are jostling each other with relative velocities of 360 m/hr, and a major meteorite causing a 1-km-wide impact crater is expected every 23 min. In this make-believe world, high-level radioactive wastes only have to be isolated from the biosphere for ~ 1 week, that is, about 10,000 min or 10,000 spans of human life.

SELECTED LITERATURE

Dating methods: Eicher, 1968; Faure, 1977; Press and Siever, 1982 (pp. 42–45); Smith 1982 (pp. 124–139, 386–409).

Earth history and the evolution of Man: McAlester, 1968; Bishop, 1978; Barraclough, 1978 (pp. 32–41); Harland *et al.*, 1982; Smith, 1982 (pp. 250–275, 363–441); McLaren, 1983; Siever *et al.*, 1983 (pp. 132–144).

Rates of change: Gretener, 1967; Gera and Jacobs, 1972 (pp. 58–123); Bürgisser and Herrnberger, 1981; Ollier, 1981 (pp. 244–256); Hsü, 1983.

CHAPTER 5

Surface Processes: Biogeochemical Aspects

I. INTRODUCTION

A large part of the Earth's land surface, in particular the low-relief areas which are of interest to us here, is covered by a heterogeneous layer of un-consolidated rock debris known as the regolith (Gr. *rhegos* = blanket, *lithos* = stone). This overlies more or less consolidated rock formations collectively designated bedrock. The regolith generally has a layered structure, with an upper layer or sequence of layers called the soil and a lower layer or sequence of layers we shall refer to as the lower regolith. Soil is the zone of biological activity in which organic matter and humus develop and react with the products of weathering. Its lower limit is a diffuse boundary marked by the deepest roots of the native perennial plants. The lower regolith in many areas is simply the transition zone between the soil and the bedrock, a zone of physical breakup (e.g., fracturing, thermal expansion–contraction and frost action) and chemical decomposition (e.g., dissolution, hydration and oxidation–reduction) of the bedrock *in situ*. In many other areas, however, the soil does not develop *in situ*, but on deposits which are quite foreign to the underlying bedrock and which indicate that quite different surface conditions pertained in the area in the not-too-distant past. These deposits include such unconsolidated materials as glacial till and moraines (from past ice ages), sand dunes and loess (from wind action), volcanic ash layers, debris from past landslides, alluvium from earlier river systems, and the like.

Many aspects of the migration of radionuclides from shallow LLW repositories (see Chapter 2, Sections II and III,B) and from surface deposits such as uranium mill tailings (see Chapter 2, Section III,D) are related to the natural processes of weathering, soil formation, groundwater flow, physical breakdown, and chemical transport in the regolith. This chapter attempts to review these processes from the point of view of radionuclide release and transport mechanisms. After general, and certainly oversimplified, background information on soil formation and weathering, the complexity of the problem is illustrated with two examples in which migration has been monitored around active or recently active sites. The long-term management of radioactive wastes in the surface environment assumes that repositories on or in the regolith will eventually become an integral part of the natural biochemical and geochemical system which makes up the surface skin of the Earth.

II. SOIL FORMATION

The bedrock formations and the various unconsolidated deposits which often cover it provide the raw materials for the development of the soil. However, since the weathering products of many different rocks and minerals are rather similar, climate is the main factor which determines its general appearance and properties. Other factors, such as biological activity, relief, and time (age), are locally important, but in spite of the complicated dynamochemical relation between these various factors, soil formation can be thought of in terms of four main processes, related to four broad climatic systems.

Podzolization. Podzolization results in soils known as pedalfers and is the dominant process in temperate humid climates, such as the climate of central Europe north of the Alps or eastern North America (Fig. 34). Such soils support dense vegetation on an upper peaty or leafy mat layer (the topsoil or O horizon), which is underlain by a leached zone (A horizon) from which Ca, Fe and Al have been removed (Fig. 34). Below the leached zone comes a zone of accumulation in which Al and Fe are redeposited (hence the soil name ped*alfer*) as hydroxides or in clay minerals. The leaching and accumulation processes are poorly understood, but are certainly a combination of inorganic (atmospheric O_2, CO_2 production) and organic–biological (formation of organic complexes, bacteria) activity.

Lateritization. Lateritization results in the lateritic soils, composed essentially of ferric and aluminium hydroxides, mainly developed in tropical humid climates. Here, there is less accumulation of organic matter (due to stronger

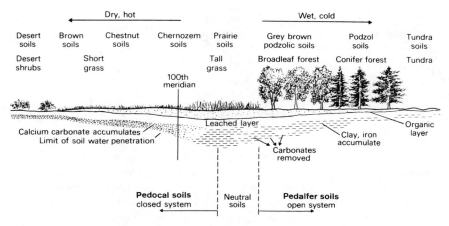

Fig. 34. Schematic diagram of the changes in soil characteristics that accompany changes in climate and vegetation between the tundra of northern Canada and the deserts of the southwestern United States. The acid soils of the humid climatic zones are known collectively as pedalfers, those of the dry climatic zones as pedocals (see Fig. 36). (From Reading, 1978, by permission of Blackwell Scientific Publications.)

oxidation) and the leaching is so intense that finally even the quartz is dissolved and removed from the soil. Lateritization is essentially an extreme form of podzolization, promoted by high rainfall and high temperatures, and both processes result in soils which are acid.

Calcification. Calcification results in soils known as pedocals and is the dominant process in semiarid to arid climates such as the western United States (Fig. 34). Characteristic of these soils is their content of $CaCO_3$ $(MgCO_3)$ due to the leaching of Ca(Mg)-bearing minerals during the rare periods of rainfall and the deposition of the otherwise soluble carbonates during the long, hot, dry periods. Such soils support only a sparse vegetation, and they contain less clay and more unaltered silicates than pedalfers. Pedocals are often neutral to weakly alkaline. The carbonate often accumulates as continuous hard layers or nodular horizons (calcrete or caliche) whose distance below the surface depends on the amount of precipitation (low rainfall produces shallow caliche).

Salinization. Salinization results in alkali soils containing various soluble salts (such as $NaCl$, Na_2SO_4, $CaSO_4$ and K_2CO_3) and is typical of very arid climates (deserts) and of poorly drained areas in other climatic zones, where evaporation rather than percolation is the main process after rainfall.

This is essentially a phenomenological classification based on the different results of otherwise poorly understood and complex physical, chemical and biological systems. We now look more closely at different parts of these systems and at the significance of some of the more general aspects for shallow land burial of LLW.

III. WEATHERING PROCESSES AND CHEMICAL TRANSPORT

Soils form by weathering, which is the general term for the alteration of unstable mineral species and the redistribution of chemical species as stable forms within the surface environment. The main agent is groundwater and the most important variables in this natural biogeochemical system appear to be the groundwater conditions, the degree of fragmentation of the regolith, the acidity and redox potential of the solutions, the properties of the insoluble residues and alteration products, and the biological activity. All man-made structures are subject to weathering, but in most cases its effects are counteracted by monitoring, maintenance and repair. Since perpetual care can hardly be guaranteed for the time periods of relevance to radwaste isolation, detailed assessment of the long-term effects of the weathering processes is required. The various factors are considered here briefly in turn (Fig. 35). Weathering processes are closely related to processes of chemical transport in the regolith (generally accomplished through the medium of groundwater), which in some situations result in dilution and dispersal of chemical species but in others cause high local concentrations (ore deposits).

A. Groundwater Conditions

The weathering processes are controlled by movement of water through the surface layers, which in turn is determined by factors such as the amount of rainfall and evaporation, the topography, and the porosity and permeability of the regolith and the bedrock. In areas of low relief, the main parameter is the depth of the water table, the surface separating the unsaturated, aereated vadose zone (above) from the saturated, unaereated phreatic zone (below). The water table is kept in approximate long-term equilibrium by the regional groundwater flow field set up in the phreatic zone by variations in the slope of the water table, which in most areas is a subdued reflection of topography. The water table is an important physical and chemical boundary which in humid low-relief areas rarely lies more than a few metres below the surface, generally within the regolith, but which in arid zones may be hundreds of metres deep, often well within the bedrock. Since most countries with extensive nuclear power programmes do not have the possibility of using arid

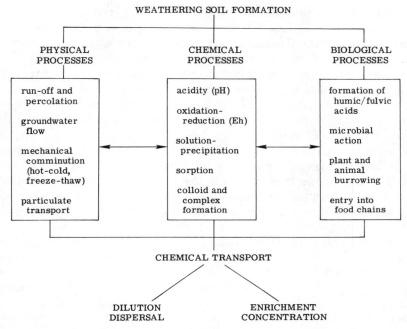

Fig. 35. Schematic representation of the factors involved in soil formation, weathering, and contaminant migration in the surface environment.

locations for LLW disposal, the problems of repository construction across and below the water table are being studied in detail. Several problems at old U.S. low-level waste disposal sites were related to disturbances of the water table (Oak Ridge, West Valley) and a detailed knowledge of groundwater conditions below uranium mill tailings piles is particularly important.

B. Degree and Type of Fragmentation

A second physical control on the system is provided by the size distributions of the fragments, particles, and mineral grains. This is partly determined by the original grain dimensions, the stratification, and the degree of fracturing of the bedrock formations or the lower regolith, partly by the form of the chemical weathering products (including insoluble residues, new minerals, and organic colloids), and partly by the mechanical breakdown of these materials during weathering. These latter processes (mechanical weathering) include cracking due to diurnal heating and cooling, frost wedging (freeze–thaw), root action, desiccation, solution, unloading, and compaction.

Such fragmentation processes have been observed at several LLW burial sites, where cracks and subsidence features provided pathways for rainwater to penetrate through the clay-rich caps into the waste trenches. Apart from affecting the water flow properties, mechanical weathering and breakdown also increase the chemical reactivity and sorption properties of the materials, since these are enhanced by increasing the reactive area available.

The water flow properties of the regolith are determined by the degree and type of fragmentation. The most common parameter for specifying the ability of a material to allow water flow is the hydraulic conductivity K, which has the dimensions of velocity (derived from the amount of water which flows through a given cross-sectional area under a given pressure head). This parameter varies from 10^{-9} cm/sec for a regolith consisting of glacial till or very clay-rich soils to 10 cm/sec for gravels and alluvial fan sediments. One selection criterion for the early LLW burial sites in the United States was low hydraulic conductivity of the host soils. This resulted in the "bathtub" effects at two sites (West Valley and Maxey Flats); water entered the trenches through breached caps and could not escape due to the low permeability. When it did escape, by overflowing at the surface or into more permeable overlying material, the leachates had had time enough to become relatively concentrated. The question of host soil permeability is undergoing a reevaluation in the siting and design of new facilities, and other hydraulic properties, such as capillary forces in partially saturated porous media, are being utilized to reduce water entry into waste trenches (capillary barriers).

C. Acidity

The main chemical factor influencing the amount of leaching of the surface materials is the acidity of the solutions. Rainwater is naturally slightly acidic due to its dissolved atmospheric CO_2 and is becoming more acidic in some areas due to the burning of fossil fuels. Groundwater is generally much more acidic because of higher CO_2 contents in the soil "atmosphere," due to *in situ* organic decay and the respiration of organisms. Mine waters and groundwaters within mine waste piles and uranium mill tailings are often strongly acidic due to the oxidation of sulphide minerals. However, reactions between fluids and soil materials in the vadose zone, together with the development of organic acids (humic, fulvic) and organic complexes, progressively change the acidity of the solutions, so that below the water table most groundwaters are neutral to slightly alkaline. The rate of solution flow and the changing acidity are among the main factors affecting the migration and transport mechanisms of the various ions, such as molecular diffusion, cation exchange, advection and dispersion. The anomalous migration effects observed around some old U.S. burial sites may partly be due to the effects of the complex and uncontrolled chemistry of the waste materials on the acidity–alkalinity (pH) of the solutions.

D. Oxidation–Reduction

The relative efficacy of oxidation (i.e., loss of electrons) and reduction (i.e., gain of electrons) in the groundwater environment, generally measured in terms of redox potential (Eh, a measure of the potential of the electron at equilibrium conditions between a redox pair), affects many biochemical reactions and influences the valence state, and hence the solubility, of the ions. This is particularly important in considering radionuclides, such as uranium or plutonium, in which one species is very soluble (e.g., U^{+6}) and the other very insoluble (e.g., U^{+4}), since the soluble species migrates much more rapidly. The redox potential of percolating waters is determined by various factors including the oxygen content of the rain, the distribution and reactivity of organic matter and bacteria, and the presence of redox "buffers" (ions with relatively rapid electron exchange properties, such as the ferric–ferrous iron couple in groundwater). The water table is again an important chemical boundary, generally marking a transition from oxidising (vadose zone) to reducing (phreatic zone) conditions, although this is by no means always the case. That groundwater tends to be reducing contributes to the retardation of elements such as uranium, which is insoluble in a reducing environment. Roll-front uranium ore deposits form due to this effect but also prove that oxidising conditions can exist in the phreatic zone over long periods of time.

E. Sorption Properties

From the point of view of the movement of ions through the system, the sorption properties of the various solid alteration products and insoluble residues are of critical importance. These are largely determined by the crystal structures of the materials, the pH of the solutions, and the state of the ions in solution (complexing). Most of the solids in soils, apart from the ubiquitous silica, are hydrated sheet silicates or hydrated oxides with a similar crystal structure. These substances are collectively called clays, a loose term covering a heterogeneous mixture of clay and claylike minerals (kaolinite, smectite–montmorillonite, illite, chlorite, vermiculite, and their many compositional and structural varieties) in aggregates with a grain size less than 2 μm. Particles of this size have a large electrical charge relative to their surface areas, resulting from imperfections or ionic substitutions within the crystal lattice or chemical dissociation reactions at the particle surface. For clay minerals in normal groundwater pH conditions, the electrical charge is generally negative and is balanced by the formation of a surface layer of positively charged ions derived from the surrounding solutions. This process of cation exchange is probably the main sorption mechanism operating in soils and results in a selective retardation of cations (including the main radionuclide species) relative to the flow velocity of the solutions involved. The parameter used to indicate cation exchange capability is the distribution coefficient K_d, an experimentally determined measure of the amount of a particular cation adsorbed by a particular solid from a particular (very dilute) solution. A more detailed discussion of laboratory-determined distribution coefficients and retardation factors and the problem of their application to the prediction of radionuclide migration rates in the field is given in Chapter 11.

F. Transport-Enhancing Mechanisms

The factors of acidity–alkalinity, redox and sorption discussed above normally combine to effect a retardation of radionuclides released to the subsurface environment under many natural conditions. There are a number of mechanisms, however, which tend to work in the opposite direction, i.e., to enhance the ease of transport up to the point that some fraction of some radionuclides may migrate at the rate of water flow. These mechanisms include complexing and transport by sorption on suspended solid particles and may include poorly understood processes of combination with large organic molecules and colloids. Complexing is the tendency for some cations to form a complex molecule by combination with charged ligands in solution, thus counteracting sorption. When the ligand is an organic molecule, the process is often referred to as chelation. It has been shown that the presence of a chelating agent (e.g., the decontamination agent EDTA, found in many low-level waste aggregates in the United States) can lead to complexing with normally less mobile radionuclides, such as ^{238}Pu, and cause them to become mobile, particularly in anoxic conditions. Particulate transport is used to designate movements in which the radionuclides are removed from solution by sorption onto, or combination with, particles which are small enough to be physically transported by the flowing medium. Field tests have shown that small but significant amounts of many radionuclides may rapidly migrate by such mechanisms even when most remain effectively immobile.

G. Biological Activity

The least understood and most complex of all the factors in weathering and soil formation is the biological activity and the organic geochemistry of decaying organic matter. The rich microfauna and microflora in the soils of every climatic zone and the aerobic and anaerobic microbial activity evidenced down to hundreds of metres below the surface affect the physical and chemical processes outlined above to an intrinsically indeterminate degree.

Also, some studies suggest that humic and fulvic acids, the main products of organic decay, can themselves form complex molecules with many metals, including some radionuclides, and hence contribute to accelerated migration.

The main conclusion from this brief survey of natural leaching and migration processes in the surface environment is that the effects of disturbing it chemically with the introduction of mixed radioactive–nonradioactive wastes in large volumes will be extremely difficult to predict. The tendency today is to strive towards a physical separation of the wastes from their environment by engineered barriers. However, the huge volumes of some waste categories (particularly uranium mill tailings) make this impractical in some cases, and some radioactive deposits are going to become effectively part of the biogeochemical surface system. An idea of the possible effects of this can be gleaned from the study of radionuclide migration around old LLW sites, as indicated in the following examples.

IV. RADIONUCLIDE MIGRATION AROUND OLD U.S. BURIAL SITES

As mentioned in Chapter 2, plutonium-contaminated solid LLW was buried in shallow trenches at at least 12 sites in the United States from 1943 onwards (Fig. 36). These sites were often badly chosen, haphazardly exploited, and poorly supervised. Three of the commercial sites were closed in the late 1970s, when it was discovered that some radionuclides were present in soils and groundwater around the sites at levels above those of normal background. Although no concentrations ever approached current radiation dose limits, the fact that radionuclides could migrate over large distances in relatively short times caused widespread concern (Table II). Since then, all sites have been subjected to close scrutiny and tight control, and the hydrogeological and geochemical data from these inadvertent field experiments is providing invaluable insights into the behaviour of radionuclides in the surface and near-surface environment under different climatic and geological conditions.

The sites can be roughly split into two groups according to climatic conditions and corresponding operating experience. The Eastern sites are all in areas of high annual rainfall ($\sim$ 100 cm average) and with shallow water tables (1–15 m). These *a priori* negative attributes were thought to be more than offset by the pedalfer soils, the regolith and/or the bedrock being rich in clay minerals (high sorption capacity, low permeability). In some sites, subsequent monitoring and exhumation tests have, in fact, proved that negligible migration of the main metallic radionuclides has taken place. However,

Table II. Content, Geological Situation, and Performance Record of Some of the Major LLW Shallow Burial Sites in the United States[a]

Site[b]	Volume LLW (m³)	Pu[c] (kg)	Regolith	Hydrology	Performance
WESTERN SITES arid climate, pedocal soils					
Hanford opened 1944	$> 2.1 \times 10^5$	30	Glaciofluvial deposits	Deep aquifer (60–115 m)	Some release (plant uptake near trench)
Idaho opened 1952	$> 1.6 \times 10^5$	370	Sands, silts, and clays	Deep aquifer (175 m) occ. flooding	Some release offsite (runoff/wind) and to depth of 80 m
Beatty opened 1962 temp. closed 1976 1979	$> 5.4 \times 10^4$	>40	Alluvial sands and gravels	Deep water table (90 m)	No release
Los Alamos opened 1952	$> 2.3 \times 10^5$	15	Weathered tuff	Deep aquifer (270–310m)	No release, except ³H near trench

EASTERN SITES
humid climate, pedalfer soils

West Valley opened 1963 closed 1975	6.7×10^4	4	Glacial till	Water table 0–7 m, "bathtub" effect	Onsite release due to overflow, some groundwater migration
Sheffield opened 1967 closed 1978	6.9×10^4	17	Glacial and glaciofluvial deposits	Water table 1–15 m	Onsite release in groundwater (^{3}H release offsite)
Maxey Flats[d] opened 1963 closed 1977	1.3×10^5	65	Weathered shale	Shallow water table, complex hydrogeology	Offsite release due to overflow and groundwater migration
Oak Ridge opened 1944	$>2.1 \times 10^5$	15	Weathered shale	Water table 1–12 m, "bathtub" effect	Offsite release due to groundwater migration
Barnwell opened 1971	$>8.5 \times 10^4$	>1	Coastal Plain alluvium	Water table 11 m	No release

[a] Main sources: Carter et al., 1979; Lipschutz, 1980; Burton et al., 1982.
[b] See Fig. 36.
[c] Estimated total plutonium content.
[d] See Fig. 37.

Fig. 36. Major LLW shallow-land burial sites in the United States with the general distribution of pedocal and pedalfer soils (as indicators of general arid and humid climatic conditions). For details, see Table II. (Data from Carroll, 1970, and Carter *et al.*, 1979.)

at other sites anomalous mobility of some radionuclides was observed. In some cases, this was caused by bad design and surface runoff (trenches filling with water and overflowing at Maxey Flats and West Valley); in others, there is no doubt that the expected retention mechanisms in the soils broke down (e.g., due to organic complexing at Oak Ridge, or due to cracks and fractures providing the water flow pathways at Maxey Flats). At the Sheffield site, the unexpected mobility was due to a lack of appreciation of the hydrogeological complexity of the enclosing medium, a formation of glaciofluvial sediments in which the permeable sand units are not isolated lenses, as originally thought, but laterally connected aquifers which allowed the off-site migration of tritiated water.

The Western sites have this in common: low annual rainfalls (10–45 cm averages), pedocal soils and very deep water tables (or water confined within a deep water-bearing aquifer). Water from the occasional rains does not percolate down to the water table in many sites; it remains trapped by capillary forces within the upper 20 m of the vadose zone and eventually evaporates. At the Nevada Test Site, plutonium from nuclear bomb testing has only penetrated 5–10 cm into the surrounding regolith over the last 10–15 years, and there is evidence that this took place by the down-washing of clay particles on which it was adsorbed. At the nearby Beatty site, there is no evidence for any leaching or transport at all. Around the Idaho site, which contains by far the biggest inventory of all the U.S. burial grounds, flooding and wind action have spread low concentrations of transuranics over an area of 2–3 km radius and percolating waters have transported them along fractures in the bedrock to about 80 m below the trenches. In general, however, there is a consensus that a carefully chosen and administered site in an arid or semiarid area with a thick, porous–permeable vadose zone provides the best guarantee for long-term LLW containment.

Example: Maxey Flats

Maxey Flats is a low, flat-topped hill in northeastern Kentucky, the United States, lying ~ 100 m above the surrounding valley bottoms (Fig. 37). The waste trenches are excavated in a regolith of weathered shale, up to 8 m thick, which grades laterally into the unconsolidated colluvium and alluvium of the hillsides and valley bottoms. Bedrock consists of Devonian–Carboniferous argillaceous rocks (80%) with many thin and some thick (up to 10 m) sandstone intercalations. The weathered shale of the burial site has a very low permeability, whereas the bedrock hydrogeology is complex and heterogeneous, dominated by the different permeabilities of the sandstone beds and the widespread jointing, which varies in intensity in the different strata. On-site migration of some radionuclides has been proved in one of the more permeable, fractured sandstones, in which ^{60}Co and ^{54}Mn have been detected up to 80 m from the nearest trench. Percolation out of most trenches, however, is so slow that water infiltrating through the trench caps (compacted shale and clay) accumulates and has to be continually pumped out and evaporated. Tritium (^{3}H) is the main radionuclide in the trench waters, but they also contain dissolved ^{90}Sr, ^{60}Co, ^{134}Cs, ^{137}Cs, ^{241}Am, and Pu isotopes, together with numerous organic waste compounds (potential complexing agents). In spite of the bedrock fracturing, a conceptual model for the groundwater flow system indicates that only 0.5% of the groundwater discharging into the surrounding streams is derived from trench water percolating into the underlying bedrock. Most enters the streams from overland runoff, and the slight contamination detected in the streams (^{3}H and ^{90}Sr in stream water, some other radionuclides in stream sediments) is thought to be mainly due to surface washing of contaminated soil from the surface of the burial site. If pumping stopped, the trenches would soon overflow, and the leachates would join the runoff and rapidly enter the drainage system. On the other hand, covering the burial site with material of very low permeability

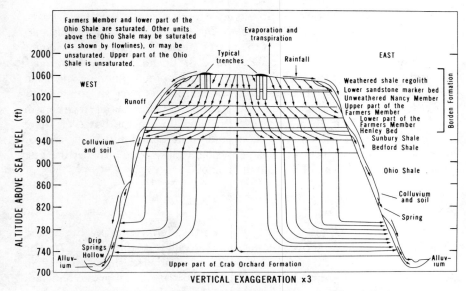

VERTICAL EXAGGERATION x3

Arrows below ground level represent flowlines. Length and density
of flowlines do not indicate velocity or volume of flow.

Fig. 37. Diagrammatic hydrogeological section of the Maxey Flats shallow land burial site, Kentucky. Groundwater flow in the bedrock takes place mainly along fractures and only contributes ~0.5% of the mean annual stream discharge. About 90% of the stream discharge is contributed by runoff and most of the waste radionuclides found in the stream sediments are thought to be due to surface transport of contaminated soil. (From Zehner, 1983.)

to reduce infiltration into the trenches would cause increased runoff and accelerated erosion of the surrounding hillsides. Additional monitoring will be necessary to assess the best way of preparing the site for decommissioning, and after remedial action, custodial care will be required for many years.

V. LONG-TERM BEHAVIOUR OF URANIUM MILL TAILINGS

The problem of the long-term stability of uranium mill tailing deposits shows some similarities to that of the shallow land burial of solid LLW from reactors and reprocessing plants. "Management" in the early postwar years was poor (see Chapter 2, Section III,D), leaving us with many examples of the behaviour of these materials under various unfavourable conditions—a collection of full-size, unintentional field experiments. Foresight and a proper understanding of geological and geochemical processes may have averted the worst of the ensuing environmental problems, but hindsight, detailed monitoring and remedial action, though expensive, have contributed enormously to our appreciation of the complexity of the systems involved. Good management is now seen to imply a planned period of surveillance and monitoring with the aim of ensuring that after this period of active involvement the tailings become an integral part of the natural surface environment, with predictable long-term stability and predictably low release of radioactive contaminants. Since this depends very strongly on the specific characteristics of the individual sites, there is no way of doing justice in a few pages to the vast amount of scientific research which has been stimulated by this problem in recent years. Also, it has become clear that the nonradioactive contaminants are of even greater concern than the radionuclides in uranium mill tailings when considering the long-term hazards. For the purposes of this book we shall simply take one representative example, after reviewing the main factors which affect the long-term geochemical behaviour of tailing deposits.

A. Uranium Ore Type

The radioactive contaminants derived from uranium mill tailings are essentially the daughter isotopes of the uranium and thorium decay series, in particular ^{230}Th, ^{226}Ra, ^{210}Bi, and the gas ^{222}Rn. Equally important, however, are the nonradioactive contaminants, which depend strongly on the uranium ore type and its associated minerals. The sandstone type (roll-front) uranium deposits of the west–central United States, for instance, show unusual concentrations of Se, Mo, As and V, and these are of at least as much concern in tailings management as the radionuclides. The Proterozoic conglomerate ores at Elliot Lake, Canada, yield significant concentrations of Ni, Co and Zn to the tailings and are very pyrite-rich (5–15%). The high pyrite content promotes an acid environment within the tailings (i.e., enhances leaching processes), and the high quartz and low clay content also has important implications for waste management (i.e., low sorption capability). In general, a detailed knowledge of the mineralogy and chemistry of the specific ores being processed is essential for developing an effective long-term strategy.

B. Processing Method

The grain size and depositional environment of the mill tailings are clearly important in assessing such aspects as seepage rates, compaction, slope stability and resistance to wind and water erosion. From the point of view of contaminants and geochemical behaviour, however, the main factor is the solution chemistry of the sludges released to the tailings ponds, which in turn depends in part on the chemical process used. For instance, the acid leach process, coupled with solvent extraction or ion exchange, yields acid solutions (pH range 1.4–1.9) with high concentrations of SO_4^{2-}, Cl^- and NH_3 and uses a range of other reagents (e.g., pyrolusite, lime, cyanide and organic compounds) which could significantly affect the long-term contaminant transport within and around the tailings. The waste sludges may also be subjected to further treatment, e.g., addition of $BaCl_2$ to coprecipitate ^{226}Ra and/or addition of lime to reduce acidity, whose long-term environmental implications also have to be assessed.

C. Impoundment System and Hydrogeology

Depending on local conditions, many different types of impoundment system are in use and various long-term stabilization procedures (close-out concepts) are envisaged. The initial deposition and dewatering of the sludges may take place in artificial tailings ponds (behind engineered dams or barriers), in natural depressions or in existing lakes (preferably with euxinic bottom conditions). Solutions from the sludges may be allowed to seep into the underlying regolith or bedrock (in amounts depending on the hydrogeological situation, the rate of evaporation, the degree of recycling, etc.), or they may be totally restricted by artificial pond linings. Pond capacity may be expanded by building new retention embankments, by increasing the height of existing dams, or by thickening existing deposits with wedge-shaped layers from raised sludge outlets. Impoundment may be permanent, envisaged as the final resting place of the tailings, or temporary, envisaged as a first stage in the backfilling of mines or in shallow land burial. All these factors affect the final size and shape of the deposits and their relation to the local geomorphological and hydrogeological system.

D. Geochemical and Biological Processes

Natural geochemical processes due to weathering and subsurface fluid circulation will take place according to the specific attributes of site geology, climate, ore type and processing. In this context, the role of climate can hardly be overemphasized. The problems of the semitropical Rum Jungle mining area of northern Australia (deep oxidation, high sporadic seepage rates during rainy seasons), of the semi-deserts of the west–central United States (minimal seepage of highly saline solutions, little seasonal variation), and of the cool temperate climate of the uranium areas in southern Canada (lower evaporation, seasonal freezing and thawing with high spring runoff) are obviously going to be completely different. Climate also directs the biological processes

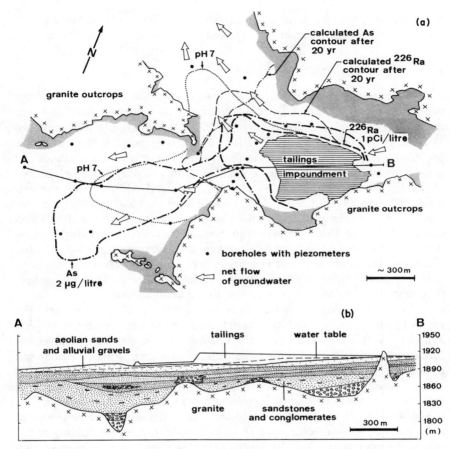

Fig. 38. Contaminant migration around the Split Rock uranium mill tailings impound-ment in central Wyoming. (a) Map of the impoundment showing the pH, As, and ^{226}Ra plumes as determined from monitoring the 30 boreholes. Also shown are the positions of the As and ^{226}Ra plumes after the next 20 years, on the basis of complex physical/geochemical modelling. (b) Cross section of the site along line AB. (Data from Taylor, in Brawner, 1980.)

necessary for soil formation and revegetation and, to a lesser extent, for the microbial activity within the tailings. This latter is known to be involved in a wide range of weathering processes, including the breakdown of silicates, the oxidation of organic matter and sulphides, and the solution and precip-itation of Fe, Mn, U, Cu, Al and other metals. The fundamental importance of microorganisms such as *Thiobacillus* and *Metallogenium* in affecting the rate of sulphide degradation (and hence the acidity) is now well established, and the possibilities for adding suitable bactericides to tailings is being investigated, a further complicating element in long-term assessment of tailings behaviour.

E. Example: Split Rock

As an example of the results of a field investigation of an active tailings deposit aimed at determining actual contaminant migration rates and predicting future developments, we take the Split Rock uranium mill in central Wyoming, the United States (Fig. 38). The tailings pond at this site is impounded behind an embankment built on a regolith of wind-blown sands and alluvial gravels, with a thin pedocal soil ($CaCO_3$ 0.1–6.3%) and a high permeability (hydraulic conductivity 1–2×10^{-2} cm/sec). Slurries from the uranium mill (typically containing 50% by weight of sand and mud) have a very low pH, due to the acid leach processing, and contain ^{226}Ra and ^{230}Th in concentrations far above the maximum permissible concentration for drinking water (100 times), as well as many other contaminants (e.g., As). Because about two-thirds of the water evaporates, even more concentrated solutions percolate down through already deposited tailings into the soil zone, where they react with the soil calcrete, raising the pH and causing the precipitation of most of the contained metallic species. As the $CaCO_3$ is consumed, the reaction zone migrates, causing a complicated geochemical evolution in space and time both above and below the normal water table. This involves co-precipitation and adsorption of radionuclides with and on metal oxides and hydroxides, adsorption on clay minerals of the regolith, reduction of dissolved concentrations by dispersion and admixture with the normal groundwater and chemical reactions due to increasingly reducing conditions at depth. At the Split Rock site, groundwater flow and water chemistry have been studied in an array of 31 boreholes from 18 to 120 m deep, most penetrating to the granite bedrock (Fig. 38). The results were then used to construct an analytical model of the subsurface geochemical and hydrogeological system. Although many of the subsurface processes are poorly understood (e.g., the reason for the development of the Ra and As plumes in different arms of the same groundwater stream) (see Fig. 38), projections for the next 20 years of active tailings impoundment showed maximum displacement of the contaminant concentration contours of tens rather than hundreds of metres.

VI. CONCLUDING REMARKS

The land surface of the Earth is covered by a thin skin of weathered rock and soil which marks the highly interactive contact zone between the geosphere, the hydrosphere and the biosphere. The geochemical and biochemical processes which have resulted in the redistribution of elements within this zone, (such as alteration, leaching and precipitation), coupled with the effects of percolating surface water and groundwater flow, have been studied in some detail but many are still poorly understood. In this chapter we have attempted to outline some of these natural processes and to indicate how they might affect the long-term stability of man-made surface deposits, such as LLW shallow burial sites and uranium mill tailing piles. Although the biogeochemical interactions are very variable and complex, the results of monitoring existing active and inactive surface repositories in many different climatic and hydrogeological situations may provide an empirical basis for improving the selection and preparation of sites to accommodate the expected large quantities of LLW in the future and for estimating their long-term behaviour.

SELECTED LITERATURE

General reviews of weathering, soil formation, ground-water flow, etc.: Carroll, 1970; Bloom, 1973; Park and MacDiarmid, 1975 (pp. 456–496); Drever, 1982 (pp. 138–199); Augustithis, 1983; Cope *et al.*, 1983 (pp. 263–378); Goudie and Pye, 1983; Réthati, 1983.

Weathering processes and contaminant migration: Means *et al.*, 1978; Carter *et al.*, 1979 (pp. 1119–1138); Freeze and Cherry, 1979 (pp. 383–462); Nishita, 1979; Silviera, 1981; IAEA, 1982b; Lutze, 1982 (pp. 745–790, 831–838); Narasimhan, 1982 (pp. 31–62); Lehman, 1983; Maynard, 1983.

Long-term behaviour of shallow land burial sites: Carter *et al.*, 1979; Lawrence Livermore Laboratory, 1979; IAEA, 1980b (Vol. 2, pp. 253–270); Cleveland and Rees, 1981; Cahill, 1982; Blasewitz *et al.*, 1983 (pp. 455–486); Robertson and Fischer, 1983; Mercier *et al.*, 1983; Zehner, 1983; Huff *et al.*, 1983.

Long-term behaviour of uranium mill tailings: Organization for Economic Cooperation and Development (OECD), 1978 (pp. 127–141, 373–398); Brawner, 1980 (pp. 205–244, 411–444); Landa, 1980; Tokunaga and Narasimhan, 1982; IAEA, 1982a; White and Delany, 1982.

CHAPTER 6

Surface Processes: Denudation and Deposition

I. INTRODUCTION

In the previous chapter, we treated the Earth's surface layer as a vast, dynamic, biogeochemical system, in which migration of radionuclides would take place largely by solution and diffusion processes within a stationary, solid framework. In this section, we take the view that the Earth's land surface is part of a vast, dynamic, physical system, in which release mechanisms largely involve the bodily movement of solid particles in transport media such as water (rivers), ice (glaciers), and air (wind). In the last chapter it became clear that physical processes (e.g., groundwater flow, particulate transport) played an important, locally dominant, role. Similarly, in considering denudation and deposition on land, the part played by chemical processes (e.g., solution transport of ions in river water) and biological activity (e.g., stabilization and destabilization of regolith by vegetation and human activity) will be continually touched upon. However, emphasis will be on classical geomorphology, the study of landforms, as a connecting link between rock disintegration (weathering and soil formation) (see Chapter 5) and production (formation of sedimentary rocks) (see Chapter 7). Since the relegation of Noah's flood to the sphere of myth in the early nineteenth century, geomorphology has been concerned with the erosive effects of rivers, glaciers, and winds, the transport of the eroded material by these agents, and the repeated deposition of this material in lakes, valleys and plains on its halting journey to the sea.

In this chapter, we first review the main processes of denudation, the wearing down of the land by mass wasting under the influence of gravity and by the erosive action of rivers and glaciers. We then consider the rate at which denudation takes place and the importance of determining these to make projections into the future for assessing the suitability of repository sites. In a final section, we look at deposition in terrestrial environments, particularly with respect to the transported lower regolith which is typical of large areas of the Earth's land surface and which is being considered in some areas as a host medium for subsurface radwaste repositories.

II. DENUDATION PROCESSES

Denudation is the general term for all processes involved in wearing down the land areas and in transporting the fragmented or dissolved material to the sea. A necessary forerunner and companion of denudation is weathering. Like weathering processes, denudation processes can be discussed in terms of separate aspects, on the understanding that in Nature these are merely parts of an interacting physical, chemical, and biological system. To a first approximation, one can distinguish between areal mass wasting and channelled erosion. An example of the first is mass movement under gravity down sloping surfaces, although some types of wind and glacier action would fall into this category. The main agent of the second is flowing water in the form of streams and rivers. We will review denudation processes under these two headings.

A. Mass Wasting

Gravity acts directly on any sloping, weathered surface to produce a general downslope movement of material which is one of the main agents of denudation. Where the slope is slight, or where the soil is closely bound and protected by a thick vegetation cover, this movement may be immeasurably slow, but at quite moderate slopes or under sparse vegetation, the effects of soil creep, rain wash, slumping and debris flowage become obvious and can be observed directly during storms and spring thaws. Steeply sloping hillsides are dominated by debris which has fallen, rolled or slipped from above, either as part of a slow continuous process (as in slump terraces, talus or rock glaciers) or as a result of single, violent events (such as landslides, mud flows, or quick-clay slides). Although water plays a role in all these processes, usually as a lubricant, there is a general lack of channelled flow. The lower slopes of most hillsides become covered with a continuous blanket of debris known as colluvium, and the upper slopes are continuously exposed to further weathering and downslope movement.

In very cold and in hot, dry climates, mass wasting is aided and abetted by moving ice (glaciers and ice sheets) and by moving air (winds, hurricanes) respectively. Wind is also an important agent in any other area of very sparse vegetation cover, for instance, in periglacial environments. The action of glaciers depends on the plucking and scouring action of debris-laden ice, in which the load is largely derived from the mechanical weathering and downslope movement of rocks in the mountains surrounding the glacier source area. Flow of glacier ice is gravity driven, but it is directed by the topographic slope of the ice surface, so that under certain conditions, glacial scouring can work up the slope of the bedrock surface. In contrast, the action of wind is not gravity driven and depends on the winnowing capacity of turbulent air vortices lifting and transporting the finest products of weathering when they are not bound by moisture or vegetation. This lifting and transporting mechanism is more important than direct scouring, which is confined to sandblasting and undercutting around the base of protuberances. All these processes of mass wasting (downslope movement, glacial scouring and wind action) are ineffectual as land sculpturing agents in comparison with, or in the absence of, channelled running water.

B. Erosion

Erosion is caused by channelled flow, usually of water in the form of streams and rivers (river erosion) but sometimes of other media (mud flows, valley glaciers). River erosion has two aspects which are complementary and often approximately in equilibrium, transport and downcutting. The river's load is provided by the processes of mass wasting, in most humid climatic zones by the downslope movement of detritus into the valley bottoms and by the melting of glacier ice. Transport then takes place by a combination of solution (dissolved load), winnowing (suspended load), and rolling, sliding or bouncing on the river bed (bed load). The amount and type of load a river can take up depends not only on the material delivered but also on the water flow velocity, which in turn is determined by factors such as water volume, bed gradient, and channel shape. The particle sizes carried in suspension or moved on the bed are related to the flow velocity by a complicated function, but the total solid load carried rises exponentially with increasing velocity. In one study, a river in the Himalayan foothills was found to carry 600–700 m^3 of solid load per km^2 of drainage area per year under normal flow conditions, but it was carrying the equivalent of 100,000 $m^3/km^2/year$ immediately after an exceptionally heavy rainstorm. In general, the transport of material by running water (rain wash, gullying and stream and river action) is discontinuous, long periods of slight effect being punctuated by short, sometimes catastrophic events of enormous carrying power, such as flash floods in desert areas, spring melting in high mountain regions, cloudburst floods and glacial surges.

This is also true of the other aspect of river erosion, down-cutting, which is mainly accomplished by the abrasive action of the suspended and bed load particles, except in areas dominated by easily dissolved rock materials, and which is accompanied by lateral or headward erosion in areas of strong relief (e.g., upstream migration of waterfalls). However, active down-cutting is only of local importance in many lowland rivers, since for most of their courses they flow across beds built of their own transported debris (alluvium), deposited when flow velocities decreased after times of flooding.

III. RATES OF DENUDATION

An assessment of local denudation processes and of present and future rates of denudation, taking into account possible tectonic activity or climatic change, is an integral part of the site selection and evaluation for any radwaste repository. Accelerated denudation due to reduction of the vegetation cover is today a well-known effect of human overexploitation of agricultural lands and forests but could also be the natural effect of gradual changes in climate. It has, in fact, already caused some release of radionuclides to the wider environment around early uranium tailings sites because of the lack of or the failure of revegetation programmes. Rare, catastrophic events such as ice- or moraine-dammed lake bursts, mud flows associated with volcanic eruptions, or unpredictable quick-clay slides in Arctic regions, can cause strong localized down-cutting and could affect the future integrity of a repository. Any of these processes, alone or in combination, working over the required isolation time could be a potential hazard, either by causing exhumation (exposure to active denudation at the surface) or by allowing encroachment (adversely changing properties or relationships which otherwise provided for containment or retardation). Effects which come under the latter head include alteration of the groundwater flow pattern, flow rates and chemistry by changing surface slopes, drainage patterns, and catchment areas; incision or removal of impermeable barriers or exhumation of soluble rock formations; reduction of the buffer zones required for the dissipation of mechanical and thermal effects; and alteration of the subsurface stress distribution (loading by new lakes, ice masses, etc.).

Estimates of the effects of these various processes in relation to the type of repository under consideration can only be carried out with reference to

specific sites. The following discussion, with two specific examples, illustrate
the dimensions of the problem.

A. Methods

Various methods have been used to estimate rates of denudation and erosion (Fig. 39). The most common way of obtaining an areal average is to
measure the sediment load (dissolved plus suspended plus bed loads) in stream
samples, together with the flow velocity over a certain period of time, and
divide this by the catchment area. This has yielded figures of ~ 0.05 m/1000
years for most of the major rivers of the world, and a world average denudation rate of 0.02–0.04 m/1000 years is now generally accepted. However,
smaller drainage basins always yield much higher values (~ 0.1 m/1000 years),

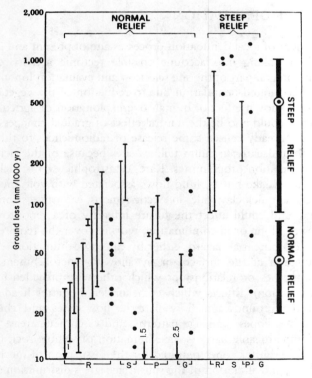

Fig. 39. Reported rates of lowering of the land surface. To the right are shown the median
values and interquartile ranges for normal and steep relief. Types of evidence: R, river load;
S, sedimentation in reservoirs; P, surface process measurements; and G, geological evidence.
(From Young, 1969, by permission of Macmillan Journals Ltd.)

exceptionally up to 13 m/1000 years in high-relief areas or in areas built of thick unconsolidated deposits. Since most river erosion occurs during exceptional floods (i.e., floods with a return period of 10 years or more), a large degree of uncertainty is attached to studies of this type if they only cover a few years. To get a more reliable long-term average, many workers have combined stream sampling with estimates of the sediment volume in lakes of known age (e.g., man-made reservoirs). This has given similar results (i.e., 0.1–1 m/1000 years), but the method suffers from not taking into account the dissolved load and from its dependence on accurate dating of the lake sediments (in the case of natural lakes). A third method involves the short-term measurement of mass movement on slopes by surveying, levelling and observation of trench walls. These techniques, alone or in combination, yield areal averages over short time periods, which are then translated into long-term figures by linear extrapolation. For instance, the minimum depth for a HLW repository in a drainage basin with a measured average denudation rate of 10 mm in 10 years would be 250 m if exhumation in the next 250,000 years is to be avoided. It is obvious that such simplistic calculations have merely an exploratory, order-of-magnitude significance.

A completely different approach is to search for geological evidence for erosion which can be coupled with dated phenomena. For example, the smooth slopes of a New Guinea volcano showed gullies averaging 0.54 m deep 30 years after eruption (rate of down-cutting: 18 m/1000 years). A 10-m-high basalt scarp in New South Wales, Australia, retreated 33 m in 35,000 years (rate of headward erosion: 1 m/1000 years). The bursting of a large glacial lake 10,000 years ago resulted in the scablands of Washington, the United States, where the bedrock was incised to a depth of 120 m in this single event. Many different kinds of evidence can contribute to such estimates and since they often refer to time periods encompassing the Pleistocene Ice Ages, they are often considered as worst case figures. However, analysis of the modification of landscapes during the Pleistocene has shown, for instance, that glacial scour in some glaciated areas has had negligible erosive effect, whereas in others many hundreds of metres of crystalline bedrock has been removed (see Chapter 13, Section IV,A). It is clear that only a balanced assessment of the results of various methods and different types of evidence in a specific area of interest can yield more than uncertain approximations. The following examples illustrate the possibilities and limitations of such assessments.

B. Example: Texas Panhandle

As a result of the screening programme to find the most favourable sites for a deep HLW repository in the Permian bedded salt formations of the southern United States, interest has focussed on the occurrences below the

Southern High Plains of the Texas Panhandle and the adjacent parts of south-eastern New Mexico (WIPP site, see Chapter 7, Section IV,B). One of the aspects studied is the past and present rate of denudation in the area, with a view to estimating its possible effects over the next 250,000 years, the nominal isolation time. The geology of the area and the results of the various methods used are summarized in Fig. 40. The retreat of the Eastern Caprock Escarpment by lateral (headward) erosion is estimated at 30–40 km over the last 250,000 years using various geological arguments. A long-term estimate of the denudation of one of the drainage basins gave a figure of 180 m over the period of its lifetime of 380,000 years, taking into account a subsidence of about 90 m due to dissolution of the underlying salt. Present denudation rates estimated from sediment load analysis vary between 0.13 and 2.23 mm/year, depending on the method used and the local conditions in the drainage basins. Roughly averaging the long-term and short-term figures gives a regional value of 100–300 m over the last 250,000 years.

Projecting into the future, it is thought that the escarpment retreat rate provides a good worst case estimate for the next 250,000 years. This is because the present climate, and the past climate as deduced from historical and archaeological data, is such as to allow maximum denudation. If the climate became more humid, the vegetation cover would increase; if it became drier, there would be much less runoff, both effects tending to reduce denudation rates. With regard to the mass wasting of the rolling plains, the long-term estimate (giving a surface lowering of 120 m in 250,000 years) is considered

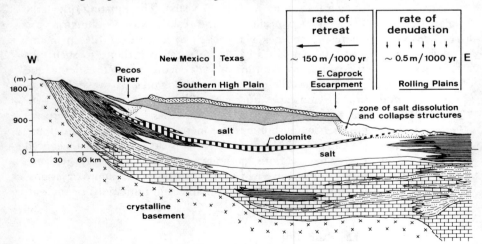

Fig. 40. Composite geological section through eastern New Mexico and the Texas Panhandle, showing the position of the Eastern Caprock escarpment and the rolling plains, with long-term erosion rates. (Data from Gustavson *et al.*, 1981.)

to be more relevant than present rates, since present denudation, already high because of optimal climatic conditions, has been sharply increased by widespread cultivation and overgrazing. Neither the retreat of the escarpment nor the general lowering of the land surface are likely to cause exhumation of a carefully sited repository in the Permian salt beds, but both may be important in significantly altering deep groundwater conditions.

C. Example: Mol, Belgium

At the nuclear fuel reprocessing plant site at Mol, Belgium, it is proposed to isolate vitrified HLW in a mined repository in clay at a depth of 200 m below the site, only 25 m above sea level (Fig. 41, and Fig. 49). Since the repository site was chosen using criteria other than its relation to possible geological release mechanisms, a retrospective study has been made as part of a safety assessment. In this, it was shown that the most important hazards in the long term (time scale 200,000 years) are likely to be sea level fluctuations and future continental glaciations. Estimates of the maximum possible effects of these processes can be made on the basis of the geological history of the North Sea basin, which is known in considerable detail for the past 60 m.y. (Tertiary and Quaternary epochs) from the results of oil and gas exploration. The maximum future range in relative sea level (determined by combinations of tectonic uplift or subsidence and changes in ocean water volume due to climatic change) lies between 150 m above and 50–100 m below its present level. Although such fluctuations could cause considerable encroachment, e.g., by coastal erosion during submergence, by headward erosion of rivers from the coastlines during emergence, or by altering the groundwater flow pattern and chemistry, a repository at a depth of 200 m is unlikely to be exhumed. The possibility of a future continental ice sheet approaching or covering the area introduces many unknowns (Fig. 41), but analysis of the effects of the last Ice Age in northern Europe suggests that depths of incision due to glacial or glaciofluvial erosion of > 100 m are not to be expected.

These examples illustrate some aspects of the problem of predicting future events and effects, which will be returned to in Chapters 13 and 14. In discussing the physical processes affecting the land surface of the earth in the present chapter, we have emphasized the aspects of removal and transport of debris. At any one moment in time, however, this debris is mainly at rest, covering most of the landscape with a blanket of unstable, unconsolidated materials temporarily deposited on their way to the sea or caught in landlocked sediment traps. We now consider these clastic deposits on land from the point of view of their possible use as enclosing media for subsurface LLW repositories.

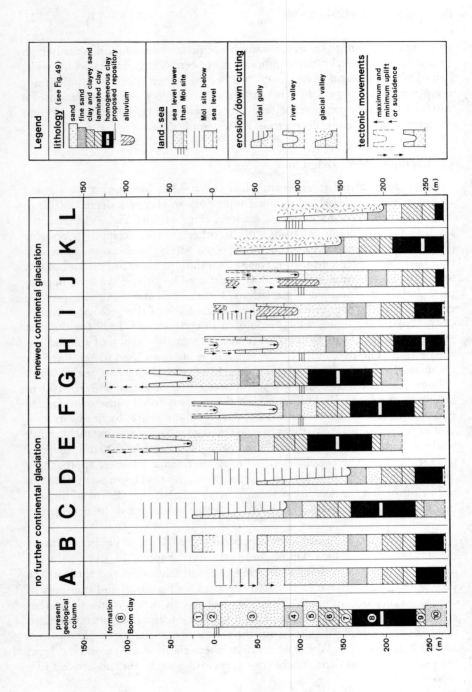

Legend

lithology (see Fig.49)
- sand
- fine sand
- clay and clayey sand
- laminated clay
- homogeneous clay
  proposed repository
- alluvium

land - sea
- sea level lower than Mol site
- Mol site below sea level

erosion/down cutting
- tidal gully
- river valley
- glacial valley

tectonic movements
- maximum and minimum uplift or subsidence

present geological column

formation ⑧ Boom clay

no further continental glaciation

A   B   C   D

renewed continental glaciation

E   F   G   H   I   J   K   L

IV. UNCONSOLIDATED TERRESTRIAL DEPOSITS AS ENCLOSING MEDIA

The unconsolidated terrestrial deposits which make up the regolith in most areas consist of the transported, fragmental products of weathering from some near or distant source area. The regolith can be roughly described as active or inactive, depending on whether the clastic deposits are the result and immediate forerunner of the surface processes at work in the area at the present time and are likely to be reworked and further transported in the immediate future, or whether the deposits are demonstrably old, the result of surface conditions not pertaining at the present time and not likely to return in the immediate future. Examples of active regolith are the alluvium of the flood plains in the lower meandering reaches of many rivers and the colluvium covering most mountain slopes. Inactive regolith is found in many regions which have been affected by continental glaciation during the Pleistocene (see Chapter 13, Section III) and now show a low relief and a thick soil and vegetation cover on top of recognizable glacial, glaciofluvial and glacioaeolian deposits such as moraines, outwash fans, eskers, kames and loess blankets. The depositional environments of the two types of unconsolidated material which are being considered as enclosing media for underground LLW repositories are illustrated in Figs. 42 and 43.

From the point of view of radwaste disposal, interest centres on inactive regolith and on the possibility of future reactivation. We have already discussed uranium mill tailing deposits, which form an artificial active regolith during the life of the mill but which must afterwards become part of an inactive natural regolith if long-term stability is to be guaranteed. Here, we consider two quite different natural situations in which inactive regolith is being considered as a possible enclosing medium for subsurface disposal of reactor and reprocessing wastes. Because the regolith is naturally unstable, i.e., unconsolidated and near the surface, even so-called inactive regolith is much more prone to adverse change under changed surface conditions than

---

Fig. 41. Synthesis of the effects of different scenarios for the future evolution of a repository in the Boom Clay, at a depth of 200 m below the Mol site, Belgium, over the next 200,000 years. Scenarios A–E assume no further continental glaciation; scenarios F–L assume renewed glaciation during a future Ice Age. (A) subsidence, climate (sea level) constant; (B) stability or subsidence, climate warm (sea level rise); (C) as (B), with tidal gullying; (D) as (A), with tidal gullying; (E) uplift, river erosion; (F) stability, glaciofluvial erosion; (G) uplift, glaciofluvial erosion; (H) subsidence, river erosion during first glacial (50,000–100,000 years); (I) subsidence, river erosion during interglacial (150,000–200,000 years); (K) glacial erosion in surge at end of first glacial; (L) glacial erosion in surge at end of second glacial. (From Vandenberghe *et al.*, in Organization of Economic Cooperation and Development, 1981b, by permission of the Organization for Economic Cooperation and Development.)

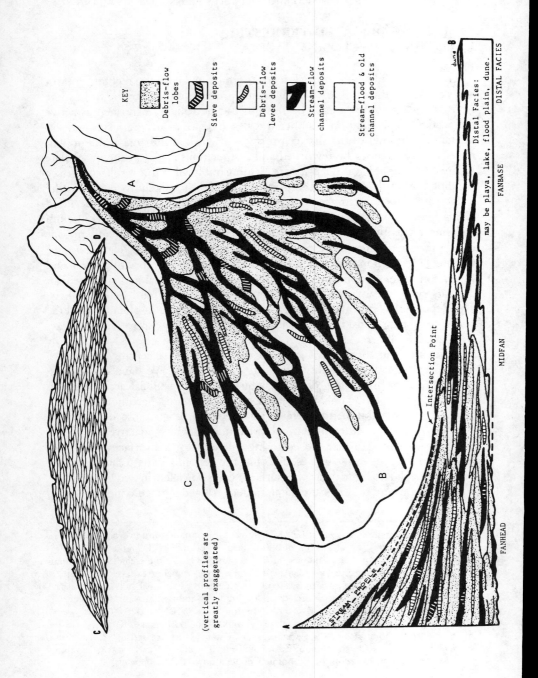

KEY

Debris-flow lobes

Sieve deposits

Debris-flow levee deposits

Stream-flow channel deposits

Stream-flood & old channel deposits

(vertical profiles are greatly exagerated)

Intersection Point

STREAM PROFILE

FANHEAD   MIDFAN   FANBASE   DISTAL FACIES

Distal Facies:
may be playa, lake, flood plain, dune.

dune B

bedrock (e.g., unconsolidated materials erode 10–1000 times faster than consolidated equivalents). For this reason, it is unsuitable for wastes requiring very long isolation times.

A. Fault Valley Fill

Fault-bounded depressions tend to become filled with detritus from the surrounding fault block mountains and fault scarps, producing large thicknesses of inactive regolith, if subsidence continues for sufficiently long periods of time (Fig. 42). Such depressions are typical of the Basin and Range province of the southwestern United States, where tensional tectonics and block faulting have been active for the last few million years (mainly postPliocene). In the area of the Nevada Test Site (Fig. 44), the fault valley fill consists mainly of fragmented sedimentary and volcanic debris washed into the depressions during the rare violent rain storms (arid climate) from surrounding bedrock mountains. In places it reaches several hundred metres in thickness. It is generally poorly stratified and poorly sorted (fragments of different sizes mixed together), and its components are poorly rounded, with marked and heterogeneous variations in the detrital lithologies. The deposits vary from sands through gravels to boulder beds, all with minor but consistently present silt and clay. Calcium carbonate (caliche) is a common cementing material at all depths, but never to the extent of producing a nonfriable rock; other secondary minerals include zeolites, cristabolite and authigenic felspars. Interspersed with these alluvial fan and debris flow deposits are occasional clay and/or evaporite horizons, representing ancient playa lakes like those typical of the present-day desert depressions.

Representative of such depressions is Yucca Flat, in the northern part of the Nevada Test Site (Fig. 44), which is underlain by up to 600 m of valley fill in places. This lies on faulted Tertiary volcanics, which in turn cover a sequence of older dolomites and limestones (Cambrian–Devonian). Interest has focussed on this area because the subsurface geology and hydrology is well known from drillholes, shafts, and craters associated with underground nuclear bomb testing. The water table lies at about 600 m below the surface, with a regional slope towards the south and west (regional flow field, see Fig. 44), and a repository near the surface would be suspended many hundreds of metres above it, in an environment which is protected from water incur-

---

Fig. 42.   Distribution of sedimentary deposits within a typical alluvial fan, with longitudinal (AB) and transverse (CD) sections. Debris flow and sieve deposits are generally gravels, whereas stream-flow and stream-flood deposits are sands which grade outwards into muds and evaporites in the distal part of the fan. (From Spearing, 1971, by permission of the Marathon Oil Co.)

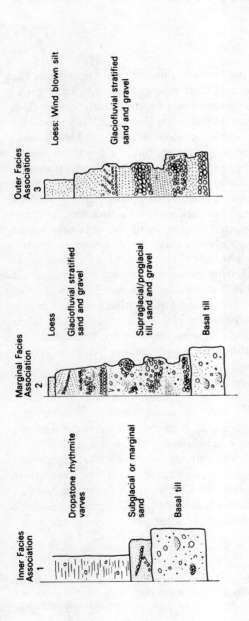

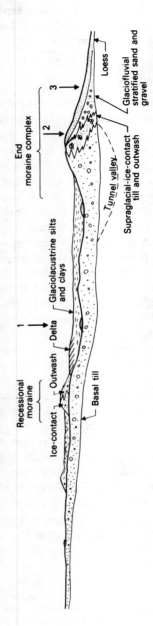

Fig. 43. Simplified sketches of the sedimentary associations (facies) which develop from the advance and retreat of an ice sheet in a terrestrial environment. The characteristic sequences may average ~ 5–50 m thick (see Fig. 45). (From Reading, 1978, by permission of Blackwell Scientific Publications.)

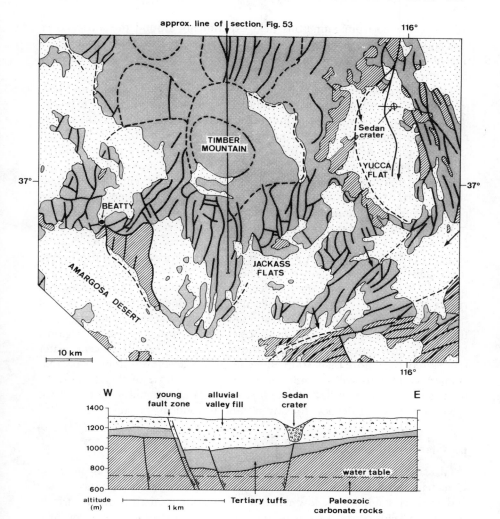

Fig. 44. Outline geological map of the Nevada Test Site and surroundings showing the distribution of deep, fault-bound, alluvium-filled desert basins. Below, a diagrammatic geological section in the region of the Sedan Crater, formed by a nuclear explosion at a depth of 200 m in the alluvium, where it is suggested that large volumes of LLW could be deposited safely. Arrows show the general groundwater flow direction in the underlying deep aquifer (Paleozoic carbonate rocks). The geological evolution of the Tertiary volcanic rocks of the region is shown in Fig. 53. (Simplified from Christiansen *et al.*, 1977, and Winograd, 1981.)

sion by the climate (most of the precipitation evaporates) and from connection to the deep carbonate aquifer by extremely low percolation velocities (estimated at 2 m/1000 years in the valley fill) and by the high sorption capacities of both the valley fill and the underlying zeolitized volcanic rocks.

All the evidence from numerous field experiments around the nuclear test sites and around the shallow LLW burial site at Beatty (see Table II) indicates that no appreciable migration of radionuclides can take place under present conditions.

The main question is then: How long are such favourable conditions likely to continue and what is the likelihood of a drastic, detrimental change before the radioactivity has decayed to innocuous levels? Most of the possible release mechanisms, such as erosion, formation of a large lake or volcanic disruption (see Chapter 8, Section III,C), can be ruled out as highly improbable on geological grounds. Climatic change in the direction of increased rainfall and lower temperatures, however, must be considered as a potentially important adverse turn of events (see Chapter 13, Section IV). Many lines of evidence point to such conditions having pertained at several times during the last 1 m.y., so-called pluvial periods related to the glacial periods in higher latitudes. However, even during the pluvials, paleobotanical evidence indicates that the climate in the Yucca Flat area was still semi-arid. Continuous sections through the valley fill (boreholes and shafts) show that the Pleistocene climate was never so humid as to prevent the development of pedocal soils (caliche horizons at all levels). Also, a large lake never developed, although periods of high influx of glacial melt waters (causing transient high water tables) cannot be ruled out. From studies of tufa deposits around springs and from regional hydrogeological considerations, it is thought that the late Pleistocene water table rarely rose more than 30 m above its present level. It is improbable that climatic conditions will become more severe than those of the Pleistocene, so sites in or on fault valley fill like that of Yucca Flat would seem to present a high degree of protection for any non-heat-producing wastes.

B. Glacial Till and Clay

Glacial deposits (Fig. 43) form an inactive regolith over wide areas of the Canadian and Baltic shields. They are protected by a thick vegetation cover and are kept stable by the low relief of the glacially scoured bedrock surface, with its myriads of small rock mounds and closed depressions. A recent Canadian study (Fig. 45) suggests that unstratified glacial clayey tills and lami-

Fig. 45. Aspects of the glacial geology of Ontario, Canada. (a) Map of the retreat of the Wisconsin ice sheet between 21,000 and 10,000 years ago, showing the major rapid readvance (surge?) about 13,000 years ago. The glacial tills in southern Canada were deposited under and around this ice margin as ground moraine or in marginal lakes (see Fig. 43). (b) The concept of emplacement of LLW canisters in large diameter boreholes in such tills, compared with an actual till log from Sarnia, Ontario, with specific characteristics indicating extremely slow groundwater movement. (Data from Denton and Hughes, 1981; Cherry *et al.*, in Carter *et al.*, 1979, and Desaulniers *et al.*, 1981.)

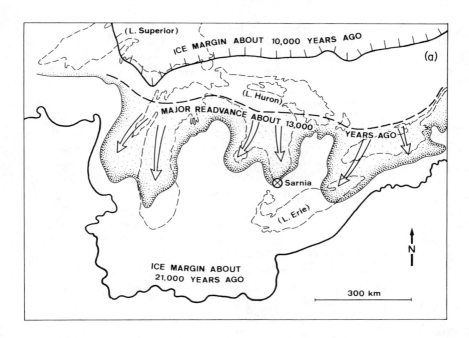

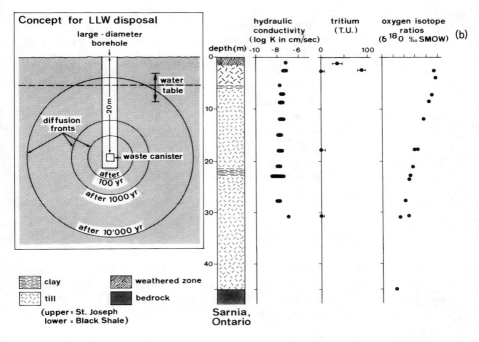

nated glaciolacustrine clays may be suitable enclosing media for subsurface LLW repositories. In the area of interest in southeastern Ontario, tills and clays cover an extensive plain to a thickness of 30–100 m. Clay minerals (illite, chlorite and minor amounts of kaolinite) and clay-sized mineral fragments make up 40–60% of the tills, with sand grains and pebbles of quartz, felspars, carbonates, shales, and crystalline rocks (bedrock components) making up the rest. An oxidised and desiccated weathered zone, 2–4 m thick and penetrated by roots, animal burrows, and desiccation or freeze–thaw fissures (containing secondary gypsum), is a common feature, but below this the deposits are unfissured and extremely impermeable. These deposits were laid down during the last Ice Age, 7500–12,500 years ago, either underneath and at the edge of the Laurentian ice sheet (see Fig. 83) or in a large freshwater lake (Lake Agassiz) along its margin.

The water table in the tills and clays of southern Canada is generally only a metre or two below the surface, rarely deeper than 5–6 m. Current regulations dictate that LLW must be solidified and deposited in subsurface burial facilities above the water table, in order to maintain them in a relatively dry state. However, poor performance of such facilities in this kind of environment, both at Canadian sites and in the eastern United States (see Table II), combined with the rarity of localities with a sufficiently thick vadose zone, has led to a reappraisal of this stipulation. Now it is suggested that LLW canisters could be emplaced in arrays of large-diameter boreholes at depths of 15–30 m *below* the water table, since the porosity (20–40%), hydraulic conductivity ($\sim 5 \times 10^{-8}$ cm/sec), and hydraulic gradients (0.01–0.1) of the clays are so low that the groundwater is essentially stagnant (flow velocities in the range 10 cm–1 m/1000 years). This has been checked at specific sites using environmental isotope ratios ($^2H/^3H$, $^{13}C/^{14}C$, and $^{16}O/^{18}O$), which showed that the groundwater still contained a considerable admixture of late Pleistocene waters. At such a site, the only release mechanism on the time frame of 1000 years (low-level reactor wastes)—assuming erosion and other external disruptive actions can be discounted—is molecular diffusion. This is a very slow and predictable process (Fig. 45) in comparison to movement in hydraulically controlled groundwater zones with complicated advective–dispersive flow patterns.

V. CONCLUDING REMARKS

In this brief review of the physical processes of mass wasting, erosion, and deposition in terrestrial environments, emphasis has been on the relevance of an understanding of these processes for the siting, design, and long-term stability of radwaste repositories. Repositories for solid LLW will tend to lie at or very near to the land surface, because of the large volumes of waste involved, and must be located and designed to withstand the direct impact of these processes over long periods of time. That this is possible in many environments is indicated by the survival of ancient man-made burial mounds, some dating back to 3500 BC. That these processes can cause rapid degeneration in other situations is proven by occurrences at several contemporary repository sites and tailings impoundments. In the case of HLW repositories, interest in denudation centres on determining the worst case for a particular area, in order to choose a repository depth which can reasonably be supposed to lie outside its sphere of influence for the required isolation time. However, even though the danger of exhumation can be eliminated, long-continued denudation can significantly affect the groundwater regime even of very deep repositories and must be taken into account. The final part of the chapter dealt with the possible use of the regolith, i.e., of unconsolidated terrestrial sediments, as an enclosing medium for solid LLW, and this leads over into the next chapter, in which sedimentary rocks are reviewed from the same standpoint.

SELECTED LITERATURE

General reviews of denudation and/or deposition on land: Bloom, 1973; Sugden and John, 1976; Reading, 1978 (pp. 15–96, 416–438); Coates, 1981 (pp. 330–529); Costa and Baker, 1981 (pp. 145–423); Laronne and Moseley, 1982.

Processes and rates of denudation: Washburn, 1967; Young, 1969; Gera and Jacobs, 1972; Lawrence Livermore Laboratory, 1980a,c; Gustavson *et al.*, 1981; Organization for Economic Cooperation and Development (OECD), 1981b (pp. 169–181); Ollier, 1981 (pp. 244–256); Zellmer, 1981; International Atomic Energy Agency (IAEA), 1982a (pp. 111–126), 1983f (pp. 93–128); Lindsey *et al.*, 1983; Sanders and Young, 1983.

Unconsolidated terrestrial deposits as enclosing media for LLW repositories: Winograd and Thordarson, 1975; Carter *et al.*, 1979 (pp. 1021–1040); Smyth *et al.*, 1979; Desaulniers *et al.*, 1981; Winograd, 1981; Perry and Montgomery, 1982 (pp. 45–55); Mercier *et al.*, 1983.

CHAPTER 7

Sedimentary Rocks: Formation and Properties

I. INTRODUCTION

The environments in which sediments are deposited on land, reviewed in the foregoing chapter, are essentially unstable, i.e., it is unlikely that such sediments will be preserved for the lengths of time and under the conditions necessary for their consolidation. Generally, terrestrial deposition is merely a temporary halt in the processes of transport from land to sea. Large areas of the continents are erosional environments, subjected to continuing uplift relative to sea level. Only when a stable depositional environment is reached, i.e., one in which, once deposited a sedimentary layer is likely to remain for long periods of time, buried under and protected from erosion by ever-increasing amounts of new sediment, will an unconsolidated sediment be preserved and gradually transformed into a consolidated sedimentary rock. Such stable environments are widespread in the world's oceans and around subsiding continental margins, although some completely terrestrial environments, those in which long-continued subsidence has been the main tectonic activity, have also provided thick rock successions at certain points in Earth history.

In this chapter, then, we start by reviewing the main environments in which sedimentary rocks are likely to form and then consider the rock-forming processes themselves, the transformation of sediment into rock which is given the general name diagenesis. From the whole spectrum of sedimentary rock types, we then choose two examples, shale and salt, for more detailed consideration. This is because both types are at present the objects of intensive study as possible host rocks for mined LLW and HLW repositories. In the final section, therefore, we review their favourable and unfavourable properties from this point of view.

II. STABLE DEPOSITIONAL ENVIRONMENTS

Present-day stable depositional environments can be subdivided into four broad groups, each with widespread past equivalents recognizable in the sedimentary rock sequences of the continents. These are inland subsiding basins, coastal alluvial plains and deltas, shallow shelf seas, and the deep ocean. An example of an inland subsiding basin, in which the sediments were typical of desert conditions, was discussed in the last chapter (see Fig. 44). Under other climatic conditions, flood plains, lakes, and deltas not unlike those of the coastal alluvial plains develop, except that marine incursions do not occur. Coastal plains and marine deltas tend to be more extensive, and rocks showing evidence of being deposited under such conditions make up a greater proportion of the geological record. Shallow marine conditions are typical of the continental shelves and some intracontinental shallow seas, and during some periods of Earth history were much more widespread than at present, covering large areas of the continental crust. They grade outwards, down the continental slopes into the deep ocean, which is perhaps the most stable depositional environment of all, with a record of 200 m.y. of continuous sedimentation in many places. Sedimentation in the deep ocean is the subject of Chapter 12. Here we concentrate on the coastal and shallow marine environments as those most relevant to the subsequent discussion of shale and salt as host rocks.

A. Coastal Alluvial Plains and Deltas

In subsiding coastal regions, major rivers flow across large thicknesses of their own detritus on extensive flood plains wherever rainfall and sediment yield are high enough to keep pace with subsidence. Complex alluvial deposits consisting of interbedded sandstones, siltstones, and shales, often with recognizable remains of river channels (channel fill gravel, point bars, etc.), old flooded soils (thin coal seams) and alluvial fans, form rock complexes many kilometres thick in places, devoid of the remains of marine organisms except in occasional interbeds representing short-lived marine incursions. Such alluvial plains are often very extensive, especially in areas where strong uplift and strong subsidence are closely adjacent, as across some major faults or at the margins of orogenic zones. When the conditions of rainfall and sediment supply, wave and tidal action, and subsidence are right, the flood plain of a major river builds outwards into the neighbouring shallow sea as a large delta, a transitional environment between continental and marine conditions. In a typical delta, the surface sediment complex of fine-grained alluvial sands and muds, with intervening swamps, levees, and bars, grades seawards into intertidal deposits and marine clays. This surface layer often lies on top of whole sequences of such sediments, some up to 8 km thick. Sedimentary rocks laid down in ancient deltas are known the world over from their association with economic coal deposits (thick coal seams being the remains of the delta swamps). When the particular conditions for delta formation are not realized, tidal estuaries, mud flats, sand bars, and the many different phenomena of coastal erosion and transport mark the transition to the marine depositional environments.

B. Shallow Shelf Seas

Marine sedimentation in shallow water is of three main types, with many possible combinations and variations. Clastic shelves and shallow seas receive mainly terrigenous material (i.e., solid particles derived from subaerial weath-

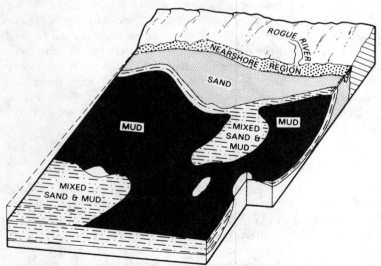

Fig. 46. Distribution of associations of clastic sediments (facies) on a typical continental margin with a schematic representation of the sandstone, mixed sand/mud (heterolithic) and mud successions. (From Reading, 1978, by permission of Blackwell Scientific Publications.)

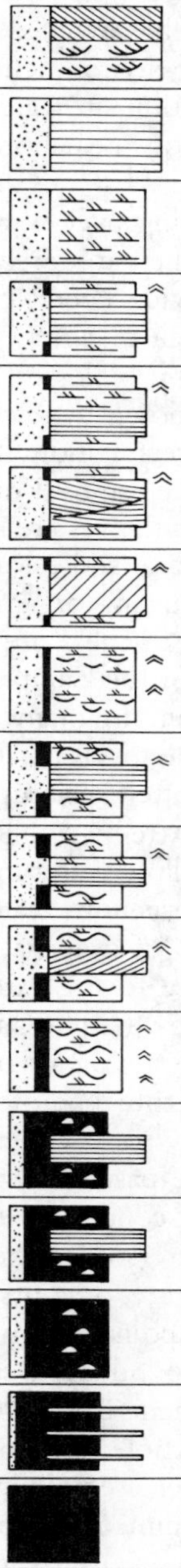

FACIES	SUBFACIES	TYPICAL LOG	INTERNAL STRUCTURE
SANDSTONE FACIES	Sa Cross-Bedded		Tabular ⎱ Cross-bedding ⎰ Trough
	Sb Flat Bedded		Parallel and low-angle lamination
	Sc Cross-laminated		Cross-lamination
HETEROLITHIC FACIES	Ha Sand Dominated		Parallel lamination
			Parallel to cross-lamination
			Low-angle and trough lamination
			Isolated tabular cross-bedding
			Sandy flaser bedding
	Hb Mixed		Parallel lamination
			Parallel to cross-lamination
			Low-angle lamination
			Flaser-wavy bedding
	Hc Mud Dominated		Parallel lamination
			Parallel to cross-lamination
			Linsen bedding
MUD FACIES	Ma		Graded sand &/or shell-rich layers
	Mb		Mud

Fig. 46. Continued.

ering and denudation) which has avoided all the other sediment traps in the alluvial, deltaic and nearshore environments. It is distributed heterogeneously by wave, tide and current action, in water depths < 200 m, but it shows a definite tendency to grade from nearshore sands and silts to offshore muds and clays (Fig. 46). Sandstone–siltstone–shale sequences deposited on subsiding clastic shelves and containing shallow-water marine fossils are well known in every geological system. They are rarely more than a few hundred metres thick, but are generally of wide lateral extent. Where the supply of land-derived clastic material is small or absent, carbonate shelves develop. Here, sedimentation is dominated by biological activity, either by the production of shell debris which breaks down to calcium carbonate sands and muds or, in some instances, by the precipitation of calcium carbonate from the sea water due, for instance, to the removal of CO_2 by organisms. Under warm climatic conditions, carbonate shelves are also associated with large reef complexes, rigid frameworks of coral and calcareous algae, filled with shell debris and bound together by encrusting organisms and precipitated calcium carbonate. Large-scale examples of these are the Great Barrier Reef of eastern Australia at the present time or the Capitan Reef (see Fig. 50) of the southern United States in the Permian. In many areas we find limestone–marl sequences with abundant marine fossils corresponding to this environment. Although relatively thin, they often span large periods of time and cover large areas, indicating periods of extreme stability.

The third main type of shallow-water marine environment is the evaporite basin. In this the influx of terrigenous material is small and biological activity is reduced because high rates of evaporation and limited water access have caused the salinity to increase to the extent that such salts as calcium carbonate ($CaCO_3$), dolomite [(Mg, Ca)CO_3], halite (NaCl), gypsum ($CaSO_4 \cdot H_2O$), and many others separate out. Although there are today some small-scale, local examples of this type of environment (e.g., playa lakes and sabka flats), many sedimentary rock sequences contain hundreds of metres of such evaporites, implying the past existence of much larger-scale and more stable equivalents. The precise conditions leading to the formation of the giant evaporite basins of the past are still controversial.

Depositional environments are, of course, much more diverse than the few mentioned above. The main point here is that in such stable situations, sediments laid down as loosely bound layers stand a good chance of being buried by successive layers of sediment belonging to the same general association and of being subjected to a whole series of physical and chemical changes as temperatures and pressures increase with increasing burial, until the sediments are transformed into solid sedimentary rock. These rock-forming processes, known collectively as diagenesis, are of some importance for understanding waste, rock and fluid interactions in deep repositories (see Chapter 11), since they take place under similar time, temperature, and pressure conditions. We now look at them in more detail with specific reference to the transformation of mud or clay into shale, and of evaporites into bedded salt.

III. DIAGENESIS

The transformation of sediment into rock is a complex interaction of physical, chemical and biological processes. These can be studied directly in deep boreholes which penetrate the more long-lived of present-day sedimentary basins (oil and gas exploration) or indirectly and phenomenologically in ancient sedimentary rock sequences now uplifted, tilted and eroded to give continuous exposures at the Earth's surface. Physical processes include compaction, fluid migration, and recrystallization due to the increasing temperatures and pressures with depth of burial. Chemical processes, such as dissolution and precipitation, which are influenced by migrating fluids of changing composition, cause cementation of the sediment grains, the main process of consolidation. Biological activity affects the chemical processes by organic decay and various bacterial reactions. Depending on the mineralogical composition, the particular combination of processes, and the depth and time of burial, the resulting rocks show different stages of transformation up to those with an intricately interlocking, nonporous fabric of crystallized and recrystallized minerals with little sign of the original sedimentary grains. This represents the extreme of diagenesis and the transition to the processes of metamorphism (see Chapter 9, Section III).

In the formation of sedimentary rocks other than evaporites, there is one simplifying factor: marine sediments tend to show a very limited mineralogy. Terrigenous input is usually the result of long-continued weathering and transport on land and consists mainly of quartz and clay minerals; biogenic input consists mainly of calcium carbonate, with varying amounts of decaying organic matter. This means that the changes such a rock undergoes during diagenesis are more directly related to the changing environmental conditions and less to the local, site-specific factors. The opposite is the case for evaporites, which show often bewildering variations in mineralogy and texture which seem to reflect the specific chemistry and climate of a particular basin rather than a general reaction to increasing depth of burial. We now look at these two situations in more detail.

A. Marine Mud Diagenesis

The main constituents of marine muds are clay minerals in varying forms and proportions. The main ones are kaolinite (a simple Al–Si form derived from tropical weathering) (typically 8–20%), illite (a more complicated Al–Si structure with K, from temperate and/or semi-arid weathering) (26–55%), smectite (a complicated Al–Si structure with Mg, Fe, Ca, Na and/or H_2O from the weathering of basic rocks) (16–53%), and chlorite (a complicated Al–Si–(Mg,Fe)–OH form, mainly derived from glacial regions) (10–18%). In addition, a number of mixed-layer clays, in which different kinds of layers alternate with each other in the crystal structure, occur, of which illite–smectite types are the most important. When muds consisting primarily of some mix of these clay minerals together with decaying organic matter are laid down in a shallow marine environment, they pass through a series of well-defined stages as they become successively more deeply buried (Fig. 47). These stages appear as a series of zones in a deep borehole through a thick sequence of marine muds and their lithified equivalents (shales).

Stage 1 is characterized by bacterial oxidation of organic matter (due to diffusion of oxygen from the over-lying seawater) and by bioturbation (disturbance by burrowing organisms), and the clays are flocculated, pelletized, and equilibriated with seawater. At depths of a few centimetres, a transition to reducing conditions (bacterial reduction of sulphate ions in trapped seawater, which releases H_2S and other products of bacterial metabolism such as CO_2, NH_3 and PO_4), marks the onset of Stage 2. In stage 2, which may

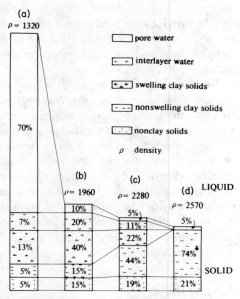

Fig. 47. The changing aspect of marine mud composition during burial, compaction and diagenesis. During diagenetic stages 1–4 most of the pore water is driven off (a–b). After this dehydration and reconstitution of the clay minerals takes place at depths of 1 to 5 km (temperatures 30–150°C). (From Leeder, 1982, by permission of the author.)

be a few decimetres to a few metres in thickness, pyrite (FeS) and calcite or dolomite are precipitated as cement or as concretions. Porosity decreases downwards to ~ 80% and new kaolinite forms in reactions between the original clay minerals, unstable carbonates, and CO_2. Ambient temperatures, however, are still those of the seafloor (a few degrees Centigrade). Below this comes the zone of bacterial fermentation reactions (possible if all sulphate is exhausted and fresh organic matter is still available), which take place at temperatures up to 30°C (depths of up to 1000 m) and which release methane (CH_4) and more CO_2. In this Stage 3, iron-rich carbonates (e.g., ankerite and siderite) precipitate, clay minerals interact continuously with the migrating pore fluids, and porosity decreases rapidly to ~ 30%. Stage 4 below is roughly marked by the temperature range 30–90°C (depths down to a few kilometres). In this zone, abiotic reactions dominate, compaction continues, and porosity decreases to < 20%. In this zone, the transition to a consolidated rock goes to completion and major mineralogical changes take place, particularly in the clay minerals (smectite → smectite/illite mixed layer → low crystallinity illite → higher crystallinity illite), releasing some ions to the pore fluids and fixing other ions transported in the pore fluids from external sources. The geochemistry of these diagenetic processes is of particular interest for radwaste disposal, because similar conditions to stage 4 are expected in the clay-rich buffer materials which are foreseen as enclosing HLW packages in mined repositories (see Chapter 10, Section V). Because of the release and upward migration of pore fluids and gases during the burial process, uplift of the rock formed in stage 4 back to the surface does not result in a reversal of any of the above reactions, so that natural analogues of the results of repository-like processes abound.

B. Evaporite Diagenesis

Evaporites show a bewildering variety of diagenetic changes, which have not yet been fitted into a consistent genetic scheme (e.g., Fig. 48). Primary minerals in evaporites (i.e., those precipitating directly from the seawater) are, in order of increasing solubility, calcite or aragonite ($CaCO_3$), gypsum ($CaSO_4 \cdot 2H_2O$), halite ($NaCl$), and various potassium and magnesium salts such as sylvite (KCl), carnallite ($KMgCl_3 \cdot 6H_2O$), kieserite ($MgSO_4 \cdot H_2O$) and bischofite ($MgCl_2 \cdot 6H_2O$). Depending on such factors as brine composition, water recharge and depth, temperature, and subsidence history, complicated reactions take place rapidly and before any significant burial. Minerals thought to be usually early diagenetic (i.e., formed before burial by reactions between primary minerals and seawater) rather than primary include anhydrite ($CaSO_4$), dolomite [($Ca,Mg)CO_3$], magnesite ($MgCO_3$) and celestite ($SrSO_4$). Both groups of minerals undergo dehydration and recrystallization and react

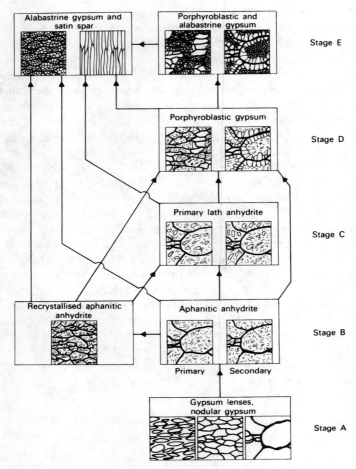

Fig. 48. Diagenetic fabrics in the Lower Purbeck evaporites of southern England, showing the complex sequence of textural changes during sedimentation and early diagenesis (stages A–B), diagenesis (stages B–C), and subsequent uplift into a near surface environment (stages C–E). (From Reading, 1978, by permission of Blackwell Scientific Publications.)

to varying degrees with migrating fluids at quite modest depths of burial. Also, further changes during uplift, including hydration, calcitization and dissolution, can cause the new growth of almost any of the above minerals, in addition to some new ones like polyhalite [$K_2CaMg(SO_4)\cdot 2H_2O$]. The result of diagenesis is a completely recrystallized rock in which the only primary feature is the original sedimentary layering in the form of fine laminations or rhythmically repeated groups of layers of different compositions. Even these may be strongly disturbed, ruptured or folded, because a further property of the noncarbonate evaporites is their low strength and high plasticity (see Chapter 10, Section III,C). Evaporites show mineral fabrics, solution phenomena, and structural relations similar to those of metamorphosed silicate rocks which have been subjected to temperatures of 600°C or more under plate boundary conditions, but many were never above 100°C and were mobilized in stable intraplate environments.

Just as an understanding of weathering processes is important for the siting and performance assessment of surface repositories, the enormous amount of data on diagenetic processes in sedimentary rocks is of direct relevance to the long-term evolution of deep-mined repositories, particularly those containing heat-producing HLW. Around them, waste, rock and fluid interactions will take place under conditions rather similar to those pertaining during diagenesis (temperature, pressure, time), particularly if the host rock is sedimentary and has not previously travelled far along the diagenetic path.

IV. SEDIMENTARY ROCKS AS ENCLOSING MEDIA

The concept of emplacing solidified HLW in deep-mined cavities or deep boreholes implies a search for the best possible enclosing medium, i.e., for a rock mass of sufficient size and sufficiently favourable properties to contain the repository completely in an environment with a high and predictable capability of retarding radionuclide migration. Most sedimentary rock formations do not fall into this category; they are generally in the form of very heterogenous, layered complexes and are often porous and/or permeable to a high degree. Two types with favourable retarding properties, however, do occur in layers of sufficient thickness and uniformity to be under serious consideration: shale and bedded salt. The term shale is used here in a wide sense to designate all rocks with a particularly high content of clay minerals (usually > 50%). It is essentially synonymous with argillaceous rocks and includes terms of more local usage, such as clay, argillite, claystone or mudstone. Bedded salt will be used for undisturbed evaporite sequences (as opposed to salt which has flowed to form diapiric structures, see Chapter 10, Section III,C) in which halite, or rock salt, is the main component but which contain minor amounts of other evaporite minerals. The shales of interest here are generally the lithified equivalents of marine muds deposited in stable shelf and shallow sea environments, whilst the bedded salt formations formed in giant evaporite basins, of which there are no present-day equivalents.

A. Shale

Shale makes up ~ 70% by volume of sedimentary rocks in the earth's crust. It generally occurs interbedded with other rock types (sandstones, siltstones, greywackes, limestones, etc.) in beds which vary from a few centimetres to a few metres in thickness. However, thick shale formations do exist which consist mainly of thick shale beds interlayered with thin, lenticular horizons of other rocks and laminae of nonclay mineral grains. Shale beds themselves are often finely laminated due to variations in the proportions of clay and accessory minerals. These small and large-scale heterogeneities represent fluctuations in factors such as sediment supply, climate and water currents in the original depositional environment. Since the time needed to accumulate a shale formation of, say, 100 m thickness is several million years (typical sedimentation rates 1–5 cm/1000 years), such heterogeneity is to be expected even under the most stable conditions. Different shale formations show different degrees of consolidation (porosity) and different degrees of anisotropy (fissility). Young shales tend to be more porous and friable than older shales because diagenesis has usually been less intense (shallow burial, short time). However, they are also less fissile, i.e., show a poorer preferred orientation of the flaky clay minerals, because the degree of compaction is less. Characterization of a particular shale formation considered as a potential host rock for a deep repository implies a detailed description of the mineralogical and lithological heterogeneity and the porosity–fissility relations at a specific site (Fig. 49).

In general, shale formations are thought to have several favourable characteristics which make them a target for extensive exploration. Foremost among these are their low permeability, their plastic behaviour under many crustal conditions and their excellent sorptive properties. General very low permeability is demonstrated by their ability to act as cap rocks for oil and gas accumulations and to maintain the abnormally high fluid pressures encountered in many oil fields, and by their role as aquiclude or aquitard in groundwater flow systems. However, these also point to possibly unfavourable attributes: their association with vitally important natural resources and their potential role in increasing fluid pressures around heat-producing radwaste (see Chapter 10, Section V). The plastic behaviour of shales is well known from their reaction to folding and thrusting in orogenic zones and is reflected in the sparsity of joints and fractures within many shale formations, particularly the younger, less consolidated ones. However, the same property creates a serious problem in young shales, i.e., the difficulty of excavating stable cavities within them. That shales (i.e., aggregates of clay minerals) show excellent ion exchange properties and other sorptive effects for the main radio-

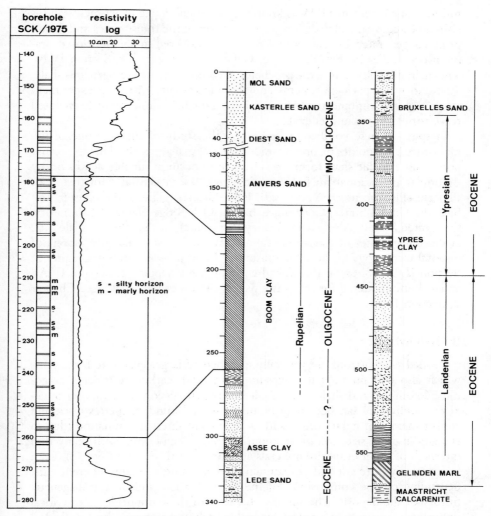

Fig. 49. Generalized stratigraphic column for the Tertiary sediments below the Mol site, Belgium, based on the results of exploratory drilling. The detailed borehole log (lithology and resistivity) to the left shows the laminated structure of the Boom Clay formation, which is being considered as a possible host rock for a HLW repository (see also Fig. 41). (From Neerdael *et al.*, in Organization for Economic Cooperation and Development, 1981c, by permission of the Organization for Economic Cooperation and Development.)

nuclides of concern in HLW is known from innumerable experiments and is also indicated by the almost complete retention of the transuranics and fission products generated by the natural reactors discovered in shale-enclosed uranium ore bodies in the Oklo mine, Gabon (see Chapter 11, Section IV,B). The main difficulty here is the geochemical complexity of heterogeneous shale, fluid and waste package systems and the intractability of *in situ* experimentation, which combine to make the long-term behaviour of shale-enclosed repositories almost unpredictable.

In spite of these conflicting properties, the study of shales is one of the main focussing points of present radwaste disposal research. This is because potentially suitable shale formations happen to occur at suitable depths below a number of national nuclear facilities [e.g., Boom clay at Mol (Fig. 49); Pliocene–Pleistocene clays at Trisaia, Italy; Eleana argillite at the Nevada Test Site, the United States; Conasauga shale at Oak Ridge National Laboratory, the United States]. Also, shale, fireclay (bentonite), or "synthetic" shale (i.e., artificial aggregates of clay minerals) are involved in several other aspects of disposal technology: as host for subsurface LLW repositories (see Chapter 2, Section II,C), as host in the seabed concept (see Chapter 12, Section IV,A), and as buffer material for the waste package in other host rocks (see Chapter 2, Section IV,1).

B. Bedded Salt

Bedded salt horizons are generally members of larger evaporite formations which also contain sulphate (gypsum, anhydrite), carbonate (calcite, aragonite, dolomite), and shale units in alternating successions and varying thicknesses. Individual salt beds may be up to several hundred metres thick and consist mainly of halite, often with a prominent internal lamination (due to changes in grain size, colour or impurity content) and often with many thin interbeds of other evaporite minerals and/or of clastic detritus (Fig. 50). Even an extremely pure salt bed can contain small amounts of many different salts, depending on the composition of the original water and on the diagenetic history of the deposit. The specific character of these impurities is of critical importance in assessing a particular bedded salt as a potential host rock. Initially, however, salt was singled out for further study on more general grounds. Foremost among these was the complete dryness of most salt mines, combined with the idea that, because halite is so easily soluble, the preservation of salt formations deposited tens or hundreds of millions of years ago should be a guarantee of stability far into the future. However, the recognition of widespread dissolution and postdiagenetic effects in many salt formations (see Fig. 40) has led to a more careful assessment of the hydrogeology of specific

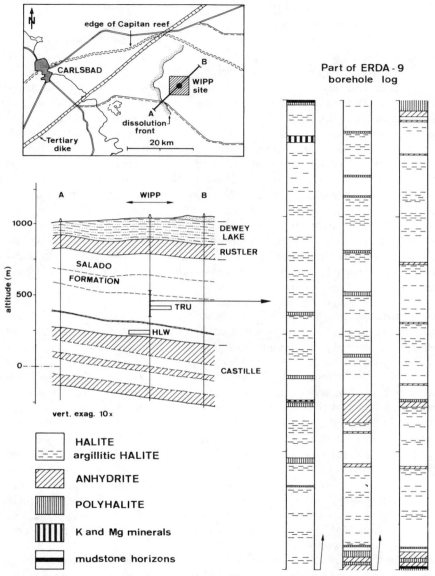

Fig. 50. Geology of the W.I.P.P. site in southeastern New Mexico, United States. An underground repository for TRU wastes and an underground test facility for HLW are at present under construction in the Permian bedded salt deposits (Salado formation) at depths of 600 to 800 m (see Figs. 15 and 40). Part of the cores from the deep borehole ERDA-9 is given on the right to show the fine lamination (laminae of clay-rich salt and other evaporite minerals), which is typical of such salt formations. (Data from Powers *et al.*, 1978.)

sites. Also, salt-mining experience has shown that although salt has a very low porosity and general lack of fractures (due to the high degree of crystal plasticity at quite shallow depths, which causes any voids to be quickly closed), salt beds often contain pockets of highly corrosive brines and, occasionally, of gases, and salt crystals themselves often contain fluid inclusions (which are known experimentally to migrate towards heat sources). Hence, the role of fluids in bedded salts is now known to be a complex matter needing detailed consideration. Other favourable characteristics, such as the ease with which underground cavities can be constructed and maintained, or the high thermal conductivity of salt (at temperatures $< 150°C$, the thermal conductivity is higher than any other rock type), have to be weighed against such unfavourable characteristics as the poor sorptive properties of halite and the association of bedded salt with important economic mineral deposits (particularly rare potassium salts). All in all, the initial "salt euphoria" has now been replaced by a more realistic and pragmatic approach, based on extensive *in situ* experimentation and predictive modelling. The particular problems associated with salt domes and diapirs are discussed further in Chapter 10.

The site of the Waste Isolation Pilot Plant in southeastern New Mexico, the United States, provides a good illustration of the geology of a bedded salt deposit (Fig. 50). The WIPP will be used to demonstrate surface and subsurface methods of handling, storing, and disposing of military wastes (TRU wastes, vitrified HLW) and spent reactor fuel and may be converted into a permanent repository for some or all of these materials, if the long-term performance assessment proves positive. Two underground cavities are projected at depths of 600 and 800 m within the Salado Formation (Permian, deposited 225 m.y. ago), which is composed of 85–90% halite, with minor interbedded anhydrite, polyhalite, and clayey–silty layers. Some horizons of potassium and magnesium salts occur, including sylvite (KCl), langbeinite $[K_2Mg_2(SO_4)_3]$ and carnallite, and in a central zone, trace quantities of these minerals are distributed throughout the halite. The minor interbeds are thin (a few centimetres) and occur in complexly alternating sequences. The thickest one, the Cowden Anhydrite (6 m), occurs near the base, where multiple anhydrite interbeds are common, except in the massive halite horizon projected for construction of the HLW cavern.

In the WIPP area, the top of the Salado Formation is marked in many places by a layer of solution breccia and residual material, and west of the site dissolution has removed, and is presently removing much of the halite from the upper part of the formation (dissolution front) (see also Fig. 40). In surrounding areas, the salt from the lower part of the Salado Formation has been sporadically removed by processes of deep differential dissolution which are poorly understood, making long-term assessment of the stability of the site extremely difficult. Related to this problem of the future effects of fluid circulation is the occurrence of polyhalite [$K_2Ca_2Mg(SO_4)_3 \cdot 2H_2O$] at regular intervals throughout the salt (Fig. 50). This mineral is commonly diagenetic (see Chapter 7, Section III,B) but here, mineral textures indicate that it formed by fluid circulation unrelated to deposition or burial, possibly up to 20 m.y. later. In general, the geology of the WIPP site shows that, in the long term, fluids must be expected to play an important but complex role in the post-diagenetic evolution of bedded salt formations.

V. CONCLUDING REMARKS

This short discussion of sedimentary rocks from the point of view of their mode of formation, emphasizing the importance an understanding of the rock-forming processes has for assessing the long-term behaviour of host rock and backfill material around mined cavities containing HLW, does no justice to the body of geological knowledge actually available in this field, based on the fact that sedimentology, sedimentary petrography, and stratigraphy are of prime importance in the exploration for and exploitation of many raw materials such as oil, coal and sedimentary ores. The main problem in using sedimentary rocks for underground disposal derives first, from this great economic interest (risk of future human intrusion) and second, from their heterogeneity (geochemical complexity) and instability at elevated temperatures. These disadvantages are less marked for the other main rock groups under consideration, volcanic rocks (Chapter 8) and crystalline complexes (Chapter 9).

SELECTED LITERATURE

General reviews of depositional environments and sedimentary rocks: Spearing, 1971; Dean and Schreibner, 1978; Reading, 1978; Hobson, 1980 (pp. 53–138); Leeder, 1982; Smith, 1982 (pp. 325–348); Herrmann, 1983 (pp. 144–159); Johnston, 1983.

Shale as repository host rock and clay as backfill: Burst, 1969; Merewether et al., 1973; Brookins, 1976; Grim and Güven, 1978; Organization for Economic Cooperation and Development (OECD), 1980b, 1981c (pp. 133–149); International Atomic Energy Agency (IAEA), 1980 (Vol. 2, pp. 41–74, 89, 104); Northrup, 1980 (pp. 403–444); Hofmann and Breslin, 1981 (pp. 16–41); Moore, 1981 (pp. 553–560); Gonzales, 1982; Van Olphen and Veniale, 1982 (pp. 771–787, 799–818).

Bedded salt as repository host rock: Powers et al., 1978; Carpenter et al., 1979; Herrmann, 1979; Anderson and Kirkland, 1980; IAEA, 1980b (Vol. 1, pp. 269–383); Northrup, 1980 (pp. 479–486); Roedder and Bassett, 1981; Weart, 1981; Gonzales, 1982; Lutze, 1982 (pp. 257–264, 439–476); Roedder and Chou, 1982.

CHAPTER 8

Volcanic Processes and Products

I. INTRODUCTION

Some of the most spectacular processes going on at the surface of the Earth are those associated with active volcanoes. Eruptions, rivers of molten rock, glowing ash clouds, fiery fountains and volcanic projectiles have awed and consternated human beings from earliest times and are immortalized in the legends and myths of ancient Greece, Scandinavia, and other areas. Compared with the slow, often imperceptible processes of weathering, denudation and sedimentation, volcanism is associated in our minds with rare, catastrophic events. Whole towns, possibly whole civilizations, have been wiped out. Whole islands have disappeared or have been born out of the sea. Zones of active volcanism are avoided when looking for sites for facilities which would only add to the catastrophy if violently disrupted, whether they be large dams, nuclear power stations, or radwaste repositories. Nevertheless, an understanding of volcanic processes and products is of considerable importance for radioactive waste disposal for several reasons. First, volcanic rocks are being intensively studied in the United States and in Japan as possible enclosing media for deep subsurface repositories. This is not because of any fundamentally favourable characteristic of such rocks, but because they happen to occur in large thicknesses below several existing nuclear facilities (e.g., Hanford Reservation, Nevada Test Site). Second, although zones of present-day or historical activity are being avoided, the question still remains in some areas whether volcanic activity will be resumed in the millenia to come and what effects such a resumption might have on a radwaste repository. Third, rapid quenching of the molten rock extruded from ancient volcanoes has often produced glasses which are not dissimilar to glasses proposed as immobilizing matrices for high-level wastes. Hence efforts are being made to elucidate the processes of alteration and devitrification of natural volcanic glasses as possible analogues for the long-term reaction of vitrified HLW under repository conditions. Within a general framework, these are the three aspects of volcanism which will be emphasized in this chapter.

II. LAVA TYPES AND ERUPTION PROCESSES

The classical picture of a volcano is the one conjured up by names like Vesuvius, Fujiyama, or Paracutin: conical mountains of ash and lava with smoking summit craters. The volcanic edifice marks a vent or chimney through the earth's crust, connecting a reservoir of molten rock (magma chamber) at depth with the surface. The edifice is built of lava (i.e., magma which has previously lost most of its dissolved gases) and tephra (i.e., rock fragments, cinders and ash ejected as solid material during eruption). The smoke consists of steam and other gases, containing ash particles, and the release of these gases from the magma as it approaches the surface is the motivating force of volcanic explosions. In contrast to most other surface processes, which are driven by solar energy, volcanism is a manifestation of the internal heat of the Earth, and as such it is usually associated with high terrestrial heat flow values (see Chapter 3, Section II,B) and with hydrothermal systems marked at the surface by hot springs, steam fields, fumaroles, and geysers.

Lava is the molten rock which issues from a volcano during eruption. It can be considered as a mixture of several metallic oxides (SiO_2, Al_2O_3, Fe_2O_3, FeO, MgO, CaO, Na_2O and K_2O), of which silica is the most abundant. Lava compositions vary between two end members: acidic lavas are high in silica (65–75% SiO_2 by weight), relatively rich in alkali metals, and poor in iron and magnesium; basic lavas are low in silica (48–58% SiO_2 by weight), poor in alkali metals and rich in iron and magnesium. The original acidic magmas at depth are thought to have been more volatile-rich (i.e., gas content 5–8%) than basic magma (gas content 2–6%). The original gas content, the viscosity and the temperature of the magma controls the eruption characteristics of the individual volcano. Acidic lava has a high viscosity, especially at temperatures near the freezing point (800–1000°C), and tends to block the vent, forming plugs, domes, and spines which are shattered explosively by the pressures building up below (e.g., Mount St. Helens). Acid volcanoes often erupt violently, ejecting huge quantities of ash and pumice as gas-charged magmatic foams, hot aerosol-like ash flows, or high plumes causing ash falls far over the surrounding country. In contrast, basic lava has a low viscosity, even near its freezing point (1000–2000°C), and tends to flow readily as sheets or rivers, the cooling, bubble-filled surface solidifying, breaking and being reincorporated in the molten lava in several characteristic ways. Basic volcanoes erupt quietly and regularly or continuously, since the dissolved gases are released quickly from the more liquid lavas with no buildup of pressures, giving the spectacular lava fountains within the craters (e.g., Hawaii). Between these two extremes, all sorts of intermediate magma–lava types occur, often showing complicated evolutionary trends in space and time, leading to many different eruption characteristics named after typical historical examples (Strombolian, Pelean, Surtseyian, Vulcanian, etc.).

Solidification of lavas produces volcanic rocks which are named according to their mineral composition. Acidic lavas solidify to rhyolite, which is generally light-coloured because of the dominance of light minerals such as quartz and felspars. Basic lavas crystallize to basalt, which is dark-coloured due to the absence of quartz and the dominance of dark ferro–magnesian minerals such as pyroxene. Both rhyolites and basalts (and the many intermediate volcanic rocks) are typically very fine-grained, often glassy due to rapid quenching, occur typically in flows a few metres to a few tens of metres thick due to the specific flow mechanisms, and show typical regular systems of open cracks due to contraction during slow, postsolidification cooling (Fig. 51). However, true rhyolites are rather rare, since acid lava has such poor flow characteristics. Most flowlike bodies of rhyolitic composition are the remains of hot ash flows or cool ash falls, which are known under the general term tuff. Ash flow tuffs show one or several welded zones due to the intense heat in the centre of the glowing avalanches, bounded by unwelded or only partially welded margins (Fig. 51), and a typical columnar fracturing, most intensely developed in the welded parts, due to differential cooling. Ash fall tuffs are typically bedded (in contrast to the unwelded parts of ash flow tuffs, which are completely structureless) and show no welded zones.

Ancient volcanic rocks of the above types occur interlayered within sedimentary rock sequences sporadically throughout the geological column, sometimes making up by far the largest proportion of the rock pile. The buried equivalents or the eroded remnants of volcanoes can be recognized in practically every geological environment. Two types of ancient volcanic activity, however, seem to have no precise present-day equivalents: acid and basic flood eruptions. Since the rock formations formed during such eruptions are the ones being intensively studied in the western United States as possible repository hosts, a short discussion of these eruption types follows.

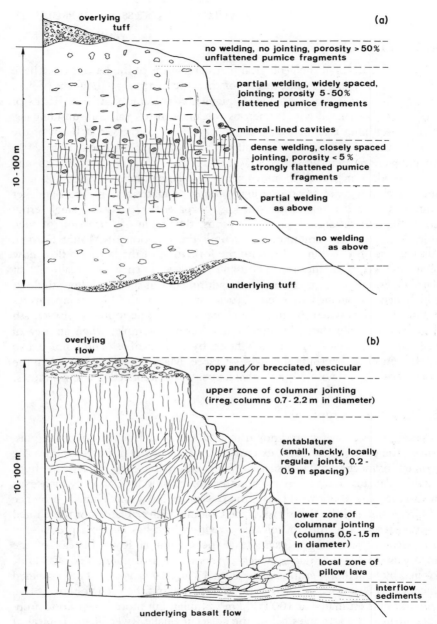

overlying tuff **(a)**

no welding, no jointing, porosity >50%
unflattened pumice fragments

partial welding, widely spaced,
jointing; porosity 5-50%
flattened pumice fragments

mineral-lined cavities

dense welding, closely spaced
jointing, porosity < 5%
strongly flattened pumice
fragments

partial welding
as above

no welding
as above

underlying tuff

10 - 100 m

overlying flow **(b)**

ropy and/or brecciated, vescicular

upper zone of columnar jointing
(irreg. columns 0.7 - 2.2 m in diameter)

entablature
(small, hackly, locally
regular joints, 0.2 -
0.9 m spacing)

lower zone of
columnar jointing
(columns 0.5 - 1.5 m
in diameter)

local zone of
pillow lava

interflow
sediments

underlying basalt flow

10 - 100 m

Fig. 51. Characteristics of volcanic flow deposits: (a) ash-flow tuffs of the Great Basin, southwestern United States (for example, on the Nevada Test Site, Fig. 44); (b) basaltic lava flows of the Columbia Plateau, northwestern United States (for example, on the Hanford Reservation, Fig. 52). (After Winograd and Thordarson, 1975, and Lehnhoff *et al.*, in Lutze, 1982.)

A. Acid Flood Eruptions

At various times in the geological past, large areas of the Earth's crust have been covered with great thicknesses of rhyolitic material in the form of piles of flowlike sheets and interbedded fragmental volcanic materials. Rocks of these types, associated with numerous volcanic centres, caldera (subsiding circular crustal plugs), and fissures, cover large areas of the Great Basin of the western United States (total area about 200,000 km^2, thickness up to 2500 m) and were poured out between 30 and 1.5 m.y. ago. Individual sheets may be hundreds of metres thick, may extend over large distances (up to 160 km), and sometimes represent staggering volumes of material (e.g., the Bandelier unit, New Mexico, represents 200 km^3 of material ejected in a single eruption). Because rhyolitic lava is so viscous, the mode of formation of such volcanic floods was not understood until the eruption of Mount Katmai, Alaska, was observed in 1912, with the formation of the large ash flow now filling the Valley of the Ten Thousand Smokes. Then it was realized that rhyolitic flood eruptions consist of the pouring out of successive hot ash flows which become welded or even partially melted by the high temperatures within the flows when the material comes to rest. The mobility of such ash flows is amply illustrated by the Mount Katmai example, when an area of 128 km^2 was covered within a few hours by 11 km^3 of ash and pumice, and by the catastrophic eruption of Mount Pelée, Martinique, in 1902 when the town of St. Pierre, with 30,000 inhabitants, was annihilated in a few minutes.

B. Basic Flood Eruptions

Because basic lavas were known to flow easily and quickly over long distances, the existence of thick, extensive basalt sheets did not present the same problem of mechanism as their rhyolitic counterparts. Nevertheless, the huge basaltic plateaus, built of many hundreds of basaltic lava flows, which exist in various parts of the world (e.g., the Deccan of India, the Thulean province of Greenland and the British Isles, the Columbia River Plateau of the United States) present a picture of volcanic activity which has no close present-day equivalent. The youngest example is the Columbia River Plateau of the northwestern United States, which covers an area of about 130,000 km^2, formed by basic flood eruptions between 17 and 14 m.y. ago. The combined thickness of lava flows exceeds 3000 m in places and the total volume of lava poured out is estimated at 100,000 km^3. Similar, but more recent flood eruptions formed the lava piles below the adjacent Snake River Plain. Typical of these and other basalt flood complexes are the systems of subvertical basaltic dikes (feeder fissures) cutting through parts of the lava piles, and the thin sedimentary interbeds, of which only a few contain appreciable amounts of volcanic ash. The nearest present-day equivalent seems to be the volcanism on Iceland, which is built almost entirely of basaltic lava flows from fissures or lines of volcanic vents associated with crustal extension along the Mid-Atlantic Ridge (see Fig. 23). The disastrous Laki (Lakagigar) eruption of 1783, the greatest outpouring of lava in historical times (volume 12.5 km³, area covered 565 km², and maximum flow distance 80 km), can be taken as a model for the type of activity, repeated thousands of times over millions of years, which built the Columbia Plateau.

III. REPOSITORY SITES IN VOLCANIC AREAS

Volcanic rock complexes of basaltic flood or rhyolitic flood type can hardly be considered *a priori* as favourable environments for radioactive waste repositories. Their main advantage lies in the fact that they immediately underlie three of the large nuclear reservations in the United States (Hanford Reservation, Idaho National Engineering Laboratory, Nevada Test Site), where political and social opposition to extensive field investigations and to repository construction and operation, is expected to be minimal. The geology of these sites seems at first sight peculiarly unsuitable (complex hydrogeology, geologically recent magmatism, and tectonism), but investigations to date indicate that some underground situations may still be found which will fulfil the necessarily more rigorous requirements of predictable, slow radionuclide release applicable to such sites. In the following, we review the volcanic rocks and history of volcanic activity at two of these sites before discussing the question of future volcanic hazards, taking the Nevada Test Site as an example.

A. Basalt as Host Rock, Hanford Reservation

The Hanford Reservation lies in the Pasco Basin of the Columbia Plateau, a depression in which the greatest known thickness of flood basalts is preserved (> 3250 m) (Fig. 52). The eroded top of the basalt pile is covered by fluvial and glaciofluvial deposits of the present and an ancestral Columbia River (0–120 m) and an older sequence of gravels, sands and clays (20–200 m) known as the Ringold Formation. These lie unconformably on the Yakima Basalt Formation (the uppermost part of the Columbia River basalts), of which the youngest preserved is 8 m.y. old. The basalts form a succession of flow layers ranging in thickness from 3 to 45 metres (with one flow, the Untanum Basalt, having a thickness of > 100 m in places). Between the different boreholes, they can be correlated on the basis of petrographic and geochemical characteristics (Fig. 52) and have proved to show constant thicknesses over distances of tens of kilometres. The basalts are grouped into members con-

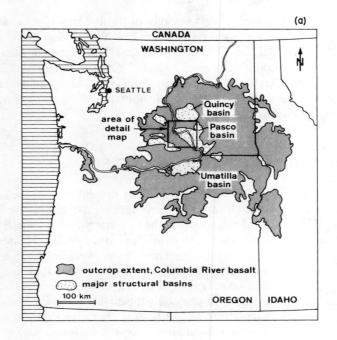

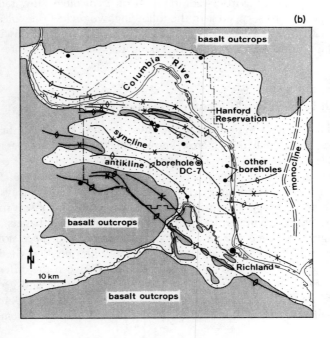

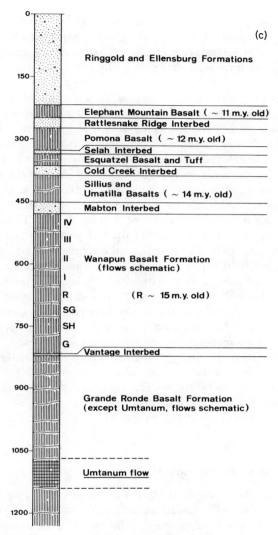

Fig. 52. Geology of the Hanford Reservation, Washington, United States: (a) location of the site in the centre of the Columbia Plateau where the basalt pile has been flexured downwards and overlying sediments preserved (Pasco Basin); (b) structural map of the Pasco Basin showing the set of open downfolds (synclines) and upfolds (anticlines) which cross the site; (c) log of borehole DC-7 showing the structure of the basalt pile and the location of the Umtanum flow. (Data from National Academy of Sciences, 1978, and other sources.)

sisting of series of flow units distinguishable only by the occurrence of vesicular, rubbly or weathered zones marking the exposed tops of flows (see Fig. 51). The various members are separated by sedimentary interbeds, 6–30 m thick, often representing water-laid sands, silts, clays, and diatomites or, more rarely, tuffs.

Interest centres at the moment on the thick flows near the base of the Yakima Basalt Formation, although the depths at which these occur on site (generally >900 m) and the resistance of compact basalt to excavation (because of its closely knit, interlocking mineral texture) will cause high construction costs. However, at this depth, most joints seem to be closed, healed, or lined with clay minerals, and the dense central parts of the flows seem to have low permeability (hydraulic conductivity ~ 10^{-8} m/sec). It is doubtful, however, whether sufficiently large volumes of rock with such favourable characteristics can be identified and used at these depths. The main problem is the heterogeneity of the system and the complexity of the deep groundwater flow pattern, controlled as it is by the scoriaceous or rubbly zones at the contacts between the flow units, by local zones of intense fracturing, by the type of the nonvolcanic interbeds (sandy = aquifer, clayey = aquitard), and by the variable leakiness of the columnar jointing. Another cause of concern is the chemical stability of the silicate minerals when subject to thermal loading, radiation and hydrothermal waste, fluid, and rock interactions, since basalt mineralogy is a metastable, very high-temperature system. Natural, low-temperature alteration products, however, and possible artificially induced ones, are mainly those which would enhance the sorptive properties of the rock mass (clay minerals, zeolites). The chemical and physical behaviour of basalt under repository conditions presently is the object of extensive laboratory testing and *in situ* investigations (Basalt Near-Surface Test Facility, Hanford Reservation).

B. Tuff as Host Rock, Nevada Test Site

At the Nevada Test Site, interest centres on the possibility of identifying a thick mass of unaltered, welded tuff (e.g., the central zone of an ash flow unit) (see Fig. 51) which is surrounded by altered tuffs containing zeolites. Most of the unwelded ash flow and ash fall (bedded) tuffs in the area show varying degrees of zeolitization due to the interaction of rhyolitic glass and high-temperature silicate minerals with groundwater. In the most highly altered tuffs, zeolites are found completely replacing the groundmass and infilling vesicles and pore spaces. It is thought that although zeolitic tuff would not itself be a favourable host rock (high porosity, low density and high content of water-bearing minerals which could dehydrate at temperatures as low as 100°C), it could act as a very effective barrier to radionuclide migration if it surrounded a mass of unaltered tuff as immediate repository host. This is because zeolites show high sorptive capacities for large radius cations and have long been in use commercially as molecular sieves, for instance, to remove radionuclides from contaminated effluents at nuclear facilities. The main problem at the Nevada Test Site is to identify such a situation at depth, since the geological environment is characterized by rapid lateral and vertical variations in rock type (due to a complex eruption process superimposed over a long period of time) (Fig. 53), widespread synvolcanic and postvolcanic faulting, and complicated patterns of hydrothermal and subaerial alteration (complex hydrogeological and geomorphological evolution). The main target area at present is Yucca Mountain, which lies south of the main area of cauldron subsidence in a region where the old ash flow tuffs are relatively undisturbed (Figs. 44 and 53).

C. Volcanic Hazards, Nevada Test Site

Siting criteria for radwaste repositories usually specifically exclude areas which have been affected by volcanic eruptions in historical times or by extensive volcanism in the last few million years. In such nonvolcanic areas, it is virtually impossible to predict whether or not volcanic activity will occur in future and, if it does, what its characteristics will be; lack of past events is taken as evidence for future quiescence. However, the Nevada Test Site has a record of comparatively recent volcanism (youngest eruptions, 270,000 years ago), which is sufficiently well dated to estimate its probability of recurrence, assuming a continuation of past trends. It is also sufficiently well exposed and researched to permit geologists to outline a plausible disruptive event and estimate its consequences. For a large repository complex (area 8 km^2) at a depth of 300 m in the Yucca Mountain area, the probability that it will be intersected by basaltic feeder pipes or fissures in the next 1 m.y. is estimated at 1:100–1:10,000, assuming that future eruption sites occur randomly within the region of known Quaternary volcanism.

What would happen if a repository were to be intersected and disrupted in this way? Field investigations, borehole logging, radiometric age dating, studies of the petrology, geochemistry and microstructure of the lavas and shallow intrusive bodies, and comparisons of the tephra cones and ash deposits with modern active volcanoes suggest the following picture. Basaltic magma formed by partial melting of the upper mantle at depths of 30–65 km will rise along deep magma conduits at velocities of 10–100 cm/sec (with pauses in one or a series of magma chambers). Although the gas content is likely to be low, magma fragmentation is expected to take place at a depth of a few hundred metres. The gases will be suddenly released, and the magma will disrupt, possibly explosively, into an upward rushing mixture of molten

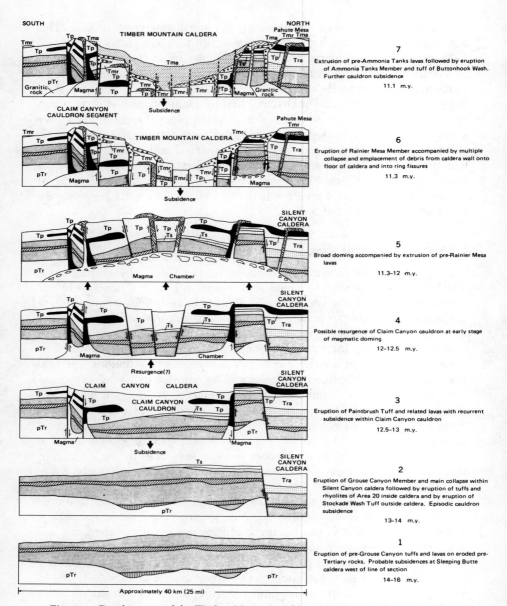

SOUTH

NORTH

7

Extrusion of pre-Ammonia Tanks lavas followed by eruption of Ammonia Tanks Member and tuff of Buttonhook Wash. Further cauldron subsidence

11.1 m.y.

6

Eruption of Rainier Mesa Member accompanied by multiple collapse and emplacement of debris from caldera wall onto floor of caldera and into ring fissures

11.3 m.y.

5

Broad doming accompanied by extrusion of pre-Rainier Mesa lavas

11.3–12 m.y.

4

Possible resurgence of Claim Canyon cauldron at early stage of magmatic doming

12–12.5 m.y.

3

Eruption of Paintbrush Tuff and related lavas with recurrent subsidence within Claim Canyon cauldron

12.5–13 m.y.

2

Eruption of Grouse Canyon Member and main collapse within Silent Canyon caldera followed by eruption of tuffs and rhyolites of Area 20 inside caldera and by eruption of Stockade Wash Tuff outside caldera. Episodic cauldron subsidence

13–14 m.y.

1

Eruption of pre-Grouse Canyon tuffs and lavas on eroded pre-Tertiary rocks. Probable subsidences at Sleeping Butte caldera west of line of section

14–16 m.y.

Fig. 53. Development of the Timber Mountain caldera over the last 16 m.y., in the area of the present Nevada Test Site. One of the localities at present under investigation as a possible site for an underground repository is Yucca Mountain (left margin of frames) (see also Fig. 44). (From Byers *et al.*, 1976.)

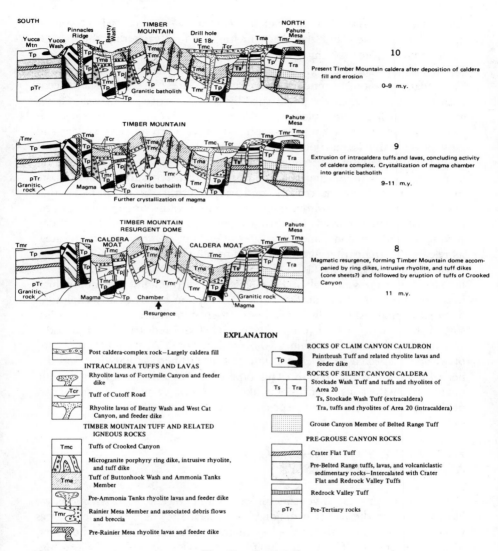

Fig. 53. Continued.

blebs, solid fragments (pumice, ash) and steam. If this takes place at or just below the repository, fragments of host rock and parts of the waste package intersected by the fissure will be torn off and rapidly transported to the surface, to be incorporated in the volcanic cone or line of cones and in the lava flows and tephra sheets associated with them. If fragmentation of the magma takes place at shallower levels than the repository, magma itself may have intruded the tunnels and chambers, melting or highly altering the various components before carrying them further.

The ensuing eruption is expected to be of Strombolian type, characterized by regular explosions of moderate intensity throwing out pasty, incandescent lava fragments called scoria, punctuated by quieter extrusion of basaltic flows and, occasionally, more violent ejection of tephra. From the wide extent of some past tephra sheets, some eruptions are expected to be more violent and ashy, possibly due to reaction of the magma with groundwater or with water bound in the zeolitic tuffs (Surtseyian eruptions). Using representative dimensions (cone size, magma volume, fissure width, repository design, waste form and rock fragment content of lavas and tuffs), it is estimated that 500–1100 m^3 of the repository contents would be transported to the surface, if such an event occurred. Of this volume, ~200 m^3 would be enclosed within the volcanic cone, and since rates of erosion are very low in this area, its release rate (redistribution in surrounding alluvial fans) is likely to be very small, perhaps about 1 m^3/10,000 years. Most of the material, however, will be enclosed in the scoria-fall and ash-fall sheets surrounding the cone, which may extend up to 12 km downwind if the eruption is particularly explosive. Such sheets have shown little resistance to erosion in the past, and it is assumed that they could be completely removed and redistributed in a wider area of alluvial valley fill within a 10,000-year period. A maximum of 20 m^3 of waste is expected to be ejected into the atmosphere as fine dust and dispersed over the whole region. This type of geological scenario, suggesting quantitatively the consequences of unlikely disruptive events, is an integral part of the risk assessment of any repository site (see Chapter 14).

IV. STABILITY OF VOLCANIC AND ARTIFICIAL GLASSES

Most of the interest in volcanic processes and volcanic rocks stems, as we have seen, from the fact that major U.S. nuclear sites happen to lie in areas where these dominate the geological environment. One aspect is, however, of more general significance: the long-term behaviour of volcanic glasses under various crustal conditions. This is because glass has become the favoured medium for HLW immobilization (see Chapter 2, Section IV,A). The short-term behaviour of vitreous waste forms of various types under a wide range of laboratory conditions is now well known after 25 years of intensive research, and the results are generally recognized to indicate high stability. However, doubts have been raised about the long term, particularly under the higher temperatures and pressures expected in deep HLW repositories. For this reason, naturally occurring glasses are coming under increasing scrutiny.

Glass is formed in nature by the rapid cooling of silicate melts. The most common process is the quenching of lava on sudden contact with air, water, or ice at the Earth's surface. Rhyolitic glass (obsidian) is much more common than basaltic glass (sideromelane or tachylite) because the high viscosity and low freezing temperature hinders crystallization, and glass is thus often a major constituent of rhyolitic lavas and ash-flow tuffs. Natural glass is also formed in situations involving strongly localized frictional heating, such as at meteorite impact sites (the main mode of formation of lunar glasses) and near crustal fractures on which there has been rapid movement (pseudotachylite). Finally, some types of meteorite are completely composed of glass (tektites), due to aerodynamic frictional heating of extraterrestrial material, and subsequent rapid cooling, during passage through the atmosphere. Since volcanic glasses are by far the most common, some aspects of their long-term behaviour as compared with vitreous waste forms are considered below. We concentrate here mainly on the processes of devitrification and leaching in aqueous solutions (see also Chapter 11, Section III).

A. Low-Temperature Alteration

A vast number of leaching experiments with simulated or actual HLW glasses, using aqueous solutions of various compositions have been carried out at low temperatures (25–90°C) and at atmospheric pressure. At the upper end of the temperature range, measured leach rates vary between 10^{-6} and 10^{-11} g/cm^2/sec, depending on glass composition, experimental setup, and fluid chemistry. At ambient temperatures, i.e., at temperatures expected in surface repositories for non-heat-producing wastes, leach rates are generally an order of magnitude lower. A long-time field experiment to study leaching under natural surface conditions is being carried out at the Chalk River nuclear facility in southern Canada (Fig. 54), where hemispheres of waste glass have been buried in fluvial sands below the water table since 1960. The average leach rate, at steady state (1966–1977), as deduced from the evolution of the ^{90}Sr and ^{137}Cs contaminant plumes in the groundwater, was 5.6×10^{-15} g/cm^2/sec, which means that it would require 20 m.y. under these conditions to completely dissolve the 12-cm-diameter hemispheres.

Archaeology provides further evidence of glass stability over long periods of time at low temperatures. Ancient glass artifacts have survived unchanged for thousands of years in wet soils and marine sediments, except for the formation of a 0.3–15-μm-thick protective layer from which Na has been leached. There is abundant evidence that such an altered, hydrated, and/or

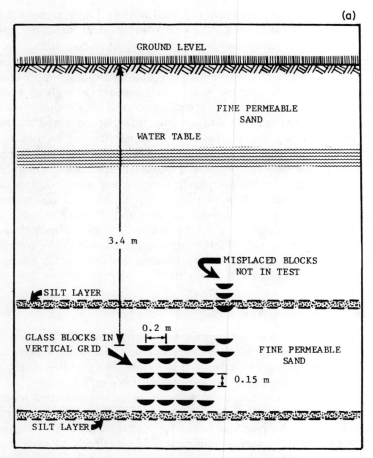

(a)

GROUND LEVEL

FINE PERMEABLE
SAND

WATER TABLE

3.4 m

MISPLACED BLOCKS
NOT IN TEST

SILT LAYER

0.2 m

GLASS BLOCKS IN
VERTICAL GRID

FINE PERMEABLE
SAND

0.15 m

SILT LAYER

Fig. 54. The low-temperature/long-time leaching experiment carried out at the Chalk River Nuclear Laboratories of Atomic Energy of Canada Ltd., Ontario, Canada: (a) arrangement of glass hemispheres (glass of nepheline syenite composition, containing ^{137}Cs, ^{90}Sr, ^{144}Ce, ^{106}Ru) in the sandy soil below the water table; (b) measured concentrations of ^{90}Sr in the soil in 1963, compared with the predictions of physicochemical modelling; (c) measured average ^{90}Sr concentration in the groundwater 1 m from the glass hemispheres at different times, compared with the predictions of physicochemical modelling.

structurally changed layer also forms on waste glasses and is responsible for the rapid drop in leach rate over the first 100–200 days of a typical leach test and over the first few years of the Canadian field experiment (Fig. 54).

Study of volcanic glasses from the point of view of quantitative estimates of long-term stability is still in its infancy. General features, however, indicate great stability under surface conditions. Volcanic glasses differ from waste

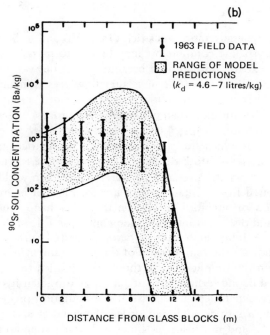

(b)

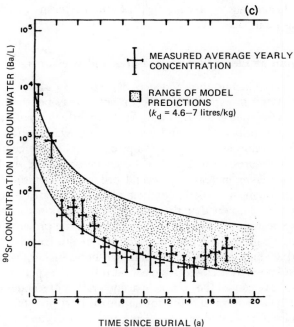

(c)

Fig. 54. Continued.

glasses in being generally more silica-rich (Table III) and in being partly crystalline. Their crystalline components may be due to partial crystallization of the lava before extrusion (mainly in basaltic glasses) but can often be shown to be due to devitrification of the glass after flow ceased (mainly in rhyolitic glasses). The oldest known rhyolitic glass was formed about 40 m.y. ago, indicating that devitrification is an extremely slow process under normal (low-temperature) conditions. Most volcanic glasses also show an extremely thin weathering rind, in which the glass is altered to a cryptocrystalline material containing clay minerals. Both devitrification and surface layer formation are known to depend on the presence of water. One 13-m.y.-old basaltic glass which was isolated from contact with water for most of its existence shows no devitrification or alteration at all. On the other hand, obsidian artifacts (used, traded, and discarded by primitive peoples for at least 30,000 yr) have been shown to exhibit, in humid climates, hydrated surface layers whose thickness depends on the temperature of the environment into which it was discarded and on the time it has lain there. This has led to the development of a widely used dating technique in archaeology (obsidian hydration dating) and to extensive experimental research in the hydration processes. It is estimated that a 20-μm-thick hydrated surface layer takes about 100 years to develop at 80°C, and much longer times at lower temperatures. These are clearly the beginnings of a wide field of study relevant to some aspects of radwaste disposal.

B. Alteration at Elevated Temperatures and Pressures

However convincing the evidence of extremely slow low-temperature leaching of glasses in experiments and in nature may be, there is some question as to its relevance to conditions in deep HLW repositories. In most, depending on depth, geology and waste loading, water-saturated conditions are likely to develop at temperatures up to 300°C in the centre of the glass cylinders at pressures up to 200 bar. Leach rates and the thicknesses of the altered surface layer increase by several orders of magnitude at these temperatures, although the experimental data base in this area is still inadequate. It has long been known that the rate of devitrification can be enormously increased by heating to moderate temperatures in the presence of water (e.g., natural rhyolitic glass millions of years old can be devitrified in the laboratory at 200–300°C) and that devitrified material shows higher leach rates than its glassy equivalent. Also, closed-system hydrothermal experiments at 300° C and 300 bar have caused the complete disintegration of standard waste glasses, with fissuring, alteration and the formation of new minerals. Although this may not in itself be a disadvantage (glass compositions can be tailored to yield alteration products which are stable and capable of retarding radionuclide

Table III. Chemical Composition of Various Radwaste Glass Matrices and Volcanic Glasses[a]

Glass	Waste load	SiO2	B2O3	Al2O3	LiO2	K2O	Na2	CaO	BaO	MgO	FeO Fe2O3	Other
Radwaste glasses												
F-SON 58.30.20	22.7	43.6	19.0	0.1	—	—	9.4	—	1.0	—	0.6	3.9
UK-209	9.8	50.9	11.1	5.1	4.0	—	8.3	—	0.4	6.3	2.7	1.1
VG 98/30	15.2	41.8	10.5	1.2	—	—	22.3	2.3	—	0.4	0.7	5.1
PNL 72-68	25.0	27.3	11.1	—	—	4.1	4.1	1.5	2.5	—	1.0	21.3[b]
ABS 39	9.0	48.5	19.1	3.1	—	—	12.9	—	—	—	5.7	1.7
GP 98/12	15.0	48.2	10.5	2.2	—	—	14.9	3.5	—	1.8	—	3.9
C31-3-EC	24.0	34.8	4.1	10.3	—	—	3.3	3.8	14.5	1.4	1.5	2.8
Ordinary window glass		70	—	1	—	—	15	9	—	3	—	2
Volcanic glasses												
Basaltic glass (sideromelane), Iceland		48.1	—	12.9	—	0.4	2.5	12.0	—	6.5	13.7	3.3
Basaltic glass, Koko Crater, Hawaii		47.5	—	14.9	—	0.9	3.7	11.6	—	6.4	11.7	2.7
Obsidian, Glass Mountain, California		74.7	—	14.1	—	4.3	4.3	1.3	—	—	1.6	0.3
World average rhyolite		72.8	—	13.3	—	4.3	3.6	1.1	—	0.4	2.6	1.9
Fresh felsitic glass, Sconser (Fig. 55)		72.6	—	13.5	—	5.3	4.1	0.5	0.1	0.1	1.2	2.6
Altered felsitic glass, Sconser (Fig. 55)		63.3	—	18.5	—	5.4	6.8	2.0	0.1	—	0.7	2.1

[a] Main sources: Dickin, 1981; Allen, in Lutze, 1982; Grauer, 1983). Numbers indicate weight percentage.
[b] Zinc oxide, ZnO.

migration), it serves to illustrate the irrelevance of standard leach tests to such conditions (see Chapter 11, Section III). As an example of the possible use of natural analogues in approaching this problem, we now look in more detail at a specific field example.

Figure 55 shows the field relations in Sconser quarry on the Isle of Skye, Scotland, where a sheet of rhyolitic glass (felsite) is cut by fractures which have been used as circulation pathways by hot aqueous fluids, before they became plugged by secondary mineral growth. The fluids were part of the Skye hydrothermal convective system, which developed above a shallow magma chamber (now represented by the Cuillin gabbro and the Redhills granite) during Tertiary times. Comparison with the hydrothermal system of the similar-sized Skaergaard intrusion of East Greenland, which is better exposed and more intensively studied, suggest the following order-of-magnitude parameters. Fluids flowed upwards in the fractures for 200,000 years, with

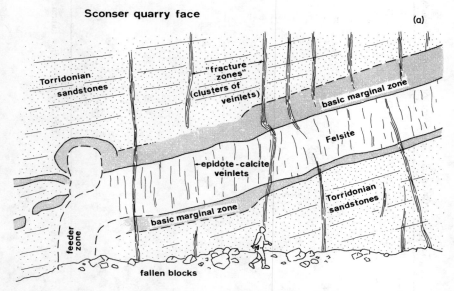

Fig. 55. Natural hydrothermal alteration of a high-silica glass. (a) Sketch of the face of Sconser Quarry, Isle of Skye, Scotland, showing the felsite sheet (high-silica volcanic glass) cutting across Torridonian Sandstone country rocks (sketch kindly supplied by A. Dickin). The sheet and country rock are crossed by vertical epidote–calcite veinlets, which were at one time fractures carrying hot aqueous solutions. (b) Chemical variations in the altered glass adjacent to the veins indicate the effects of prolonged hydrothermal leaching at temperatures of around 400°C. (From Dickin, 1981, by permission of Macmillan Journals Ltd.)

distance from edge of vein

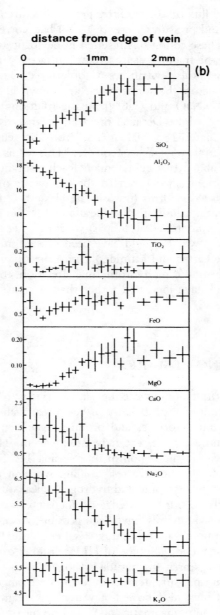

Fig. 55. Continued.

an integrated fluid flux of 10,000 litres per cm fracture length. Temperatures were ~ 400°C, and pressures were 100–300 bar, corresponding to a depth of about 1 km. These conditions resulted in devitrification and alteration of the glass forming the fracture walls across a surface layer 1.5 mm thick, outside which isochemical devitrification took place. The mineralogical and chemical changes across the surface layer are shown in Fig. 55. These show addition of Al_2O_3, NaO and CaO from the solutions and concomitant removal of SiO_2 and MgO. Detailed microprobe analysis indicates that there was a net loss of material of 0.019 g/cm^2 from the vein wall, giving a long-term average leach rate of 10^{-14} $g/cm^2/sec$. This seems to indicate that high-silica glass (see Table III) would remain leach-resistant under extreme hydrothermal conditions, in contrast to the behaviour of the presently most favoured vitreous waste forms (borosilicate glasses). Other geochemical and isotopic data from Skye indicate that the mineral phases formed in the surface layer (particularly epidote, alkali felspar) would efficiently incorporate many of the radionuclides which would have been present in a waste glass (particularly Sr, Ba, rare earths and actinides). Since the rhyolite glass has retained its structural integrity during hydrothermal alteration, the alteration products in this case would have increased rather than decreased its retention capabilities.

V. CONCLUDING REMARKS

Although a first rule of radwaste repository siting is to avoid active volcanic areas or areas with a high probability of being reactivated in the near future, the study of volcanic processes and products has been stimulated by their relevance to certain aspects of the radwaste problem. The intensive research at present underway in the region of the Nevada Test Site and the Hanford Reservation, involving extensive test drilling, work in underground laboratories and detailed petrographic and hydrogeological studies, is likely to contribute significantly to our knowledge of ancient volcanic activity in a way which would otherwise hardly have been possible. Also, the use of glass matrices for HLW immobilization has now been established as the favoured procedure, at least for the initial phase of HLW disposal early next century, and this is stimulating a search for natural analogues with well-documented stability in the relevant pressure, temperature and time range. Volcanically active areas (and related high heat flow, hot water and steam fields) are generally the surface expression of magmatic and metamorphic processes at depth; these are the subject of the next chapter.

SELECTED LITERATURE

Volcanic processes, products and hazards: Smith, 1960; Macdonald, 1972; Bolt *et al.*, 1975 (pp.
   63–131); Heiken, 1979; Sheets and Grayson, 1979; Johnpeer *et al.*, 1981; Chapin and
   Elston, 1979; Crowe and Amos, 1982; Crowe *et al.*, 1982, 1983.
Basalt as repository host rock (e.g. Columbia Plateau, northwestern U.S.A.): La Sala and Doty,
   1971; Rockwell Hanford Operations, 1977; National Academy of Sciences (NAS), 1978;
   Doctor, 1980; Moore, 1981 (pp. 51–58); Petrie *et al.*, 1981; Gonzales, 1982; Lutze,
   1982 (pp. 597–609); Reidel, 1983.
Tuff as repository host rock (e.g. Great Basin, western U.S.A.): Winograd and Thordarson, 1975;
   Byers *et al.*, 1976; Christiansen *et al.*, 1977; Johnstone and Wolfsberg, 1980; U.S. Department of Energy (DOE), 1980; Gonzales, 1982; Waddell, 1982.
Long-term behaviour of volcanic and artificial glasses: Lofgren, 1971; Bradley, 1978; McCarthy
   et al., 1978; Simmons *et al.*, 1979; McCarthy, 1979 (pp. 57–168); Grover, 1980; Karkhanis *et al.*, 1980; Northrup, 1980 (pp. 85–238); Westsik and Peters, 1980; Dickin,
   1981; Moore, 1981 (pp. 83–122, 283–290, 315–322); Lutze, 1982 (pp. 1–298); Roy,
   1982 (pp. 148–183); Grauer, 1983; Melnyk *et al.*, 1983.

CHAPTER 9

Natural and Synthetic Crystalline Rocks

I. INTRODUCTION

The preceding chapters have been mainly concerned with processes going on at the surface of the earth. "Surface" was not taken in its geometric sense, but rather in the sense of a surface layer or skin consisting primarily of unconsolidated materials. However, it was necessary to make excursions into the region below the skin at several points. The last chapter was particularly concerned with the internal energy of the Earth and with the solidified, often crystallized, products of molten material derived from the Earth's interior. We now shift emphasis to this realm of internal processes and to the formation of coarse-grained crystalline rocks at depth within the Earth's crust (see Chapter 3, Section IV). In this chapter, we are concerned with the rocks themselves, how they form and what makes them of interest to the problem of radwaste. In subsequent chapters, we will look at the uppermost crust, which is mainly composed of crystalline rocks, from the point of view of present-day internal processes and their perturbation through the insertion of a deep HLW repository. Because the transition from external processes (Chapters 5–8) to internal processes (Chapers 8–11) involves an important change in methodology, we insert here a short preamble and review.

II. STUDY OF INTERNAL PROCESSES

Internal processes, in contrast to many of the external processes discussed in previous chapters, cannot be studied directly. They must be approached more indirectly, by physical probing (borehole logging or mine geology), by geophysical techniques (such as interpretation of seismic records, electrical conductivity and magnetic anomalies), by observation of the surface expression of present-day deep activity (uplift and subsidence, plate motion and volcanism), and by describing and interpreting the exhumed remnants of rock bodies thought to have been produced by similar processes in the past (geology of crystalline rock complexes). It is the emphasis placed on this latter activity which distinguishes the study of internal processes methodologically. Theoretically, if the end result of a complex system of unknown or poorly known processes is studied in sufficient detail, the general outline of the most probable scenario or evolutionary model will be distinguishable among the many possibilities. The philosophy is the same as in reconstructing a crime from meagre clues and circumstantial evidence. The geology of a crystalline complex may be well known (geology here taken to include descriptive activities on all levels, including field mapping, microscopic study and rock and mineral analysis), all clues and evidence collected, but it may still be difficult to choose between different modes of formation. In this situation, help can only come from two directions: experiment and theory. Our knowledge and understanding of many deep crustal processes is thus based on a synthesis of geological data, experimental results and theoretical analyses (see Chapter 3). Although this book concentrates on the geological data, without a balanced integration of the other types of information, a discussion of internal processes would not be possible. This point is emphasized also because long-term predictions of repository behaviour rest on a similar philosophical foundation (see Chapter 14).

Internal processes form a connecting link among the following three chapters. In this chapter, we start by considering the processes of formation of crystalline rocks in the Earth's crust (see Chapter 3, Section IV,1), either by crystallization of rock melts (magmatism) or by the solid-state recrystallization of already-formed rocks under the action of heat, pressure, deformation and/or migrating fluids (metamorphism or metasomatism). Rocks formed in these ways are being considered in many countries as possible hosts for deep repositories. Also, ceramic waste forms, which are being suggested as alternatives to glass for HLW immobilization, are the synthetic equivalents of natural crystalline rocks and naturally occurring minerals. In contrast to volcanic glass, the stability, alteration and long-term behaviour of rocks and minerals in many different crustal situations has been a main concern of crystalline geology and this gives a possible basis for judging the future behaviour of synthetic analogues. As repository hosts, crystalline rocks now find themselves in an environment quite different to that under which they formed. They are now part of a cold, brittle basement complex within a few kilometres, generally a few hundred metres, of the Earth's surface. The present-day physical state of this uppermost part of the crust forms the theme of Chapter 10, since this is critical for all aspects of deep repository siting and design. Subjects under this head include thermal and mechanical aspects (temperature, pressure and stress distribution), fluid circulation patterns (deep groundwater flow) and structural state (fracture patterns, diapiric structures), and the various interactions among these parameters, particularly under the disturbing influences of cavity construction and subsequent HLW emplacement. In Chapter 11 we then turn to possible release and/or retention mechanisms from and/or within such deep repositories. Within the complex physical framework described in Chapter 10, radionuclide migration will depend on the geochemistry of the waste packaging, host rock and groundwater system (near-field processes) and on fluid–rock interactions along subsequent flow paths (far-field processes). It is clear that the principles discussed in Chapters 10 and 11 are not host rock-specific; the choice of examples will illustrate the range of problems associated with particular host rocks and geological environments.

III. NATURAL CRYSTALLINE ROCKS

Crystalline rock is an informal designation for all silicate rocks, and some other rock types, which show a coarse-grained interlocking fabric of crystalline phases and which show evidence of having formed at high temperatures and pressures deep within the Earth's crust. Their existence at or near the Earth's surface at the present time (i.e., at temperatures and pressures much lower than those under which they formed) is due to the subsequent uplift of the crustal segment in which they formed and to concomitant denudation (see Chapter 6, Section II). Interest in crystalline rocks as enclosing media for deep underground repositories derives from the facts that they make up a large part of the basement in all areas of the continents and that they are often very homogeneous. This applies particularly to magmatic rock bodies, such as granitic plutons, which are very widespread. The rocks themselves are tough (i.e., resistant to excavation) but strong (i.e., mechanically stable around large openings). They are also very resistant to chemical change, as evidenced by the existence of some crystalline rocks in a completely fresh state after a billion years or more in a near-surface environment (Precambrian shields). This is related to their coarse crystallinity and their extremely low porosity (often < 1%), in turn related to the interlocking nature of the mineral aggregates. The main unfavourable property of crystalline rocks with regard to radwaste disposal, their brittleness in response to crustal movements under near-surface conditions and thus their propensity to be strongly fractured, will be treated in more detail in Chapter 10. Here we consider the actual processes of formation of the rocks themselves at great depth in the Earth's crust.

A. Rock-Forming Processes at Depth

With a few exceptions, crystalline rocks are the result of ancient plate boundary activity. A review of the structure and evolution of the continental crust has been given in Chapter 3; the aim of this section is to concretize some of the concepts presented there. A typical crystalline basement complex (Fig. 56) consists of three main elements. The oldest is normally a metamorphic complex, typically heterogeneous on all scales and showing intense folding (heterogeneous flow in the solid state). Cutting discordantly through this complex, we usually find large, irregular, internally homogeneous rock bodies called plutons, which seem to have formed by the intrusion of magma into the already consolidated metamorphic rocks. The youngest elements are sheetlike rock units cutting through both the old metamorphics and the plutons, called sills, dikes or veins depending on their specific character. These minor intrusions show features indicating various modes of origin, from intrusion as magma to alteration of the host rock by migrating fluids.

Plutons are made up of coarsely crystalline rocks in which the main mineral components, in granite, quartz, felspar and mica, in gabbro, pyroxene and

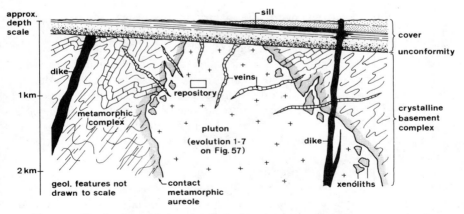

Fig. 56. Sketch of a vertical section through a typical crystalline basement complex.

felspar), are easily distinguishable by the naked eye and are homogeneously aggregated. Under the microscope, the shape and arrangement of the grain boundaries (texture) are similar to those obtained by cooling silicate melts in the laboratory and indicate the same crystallization sequences. Comparison of experimental, petrographical, and geochemical data leaves no doubt that most plutons are magmatic (as opposed to metasomatic), and allows detailed interpretation of age of intrusion, emplacement conditions, cooling history and wall rock assimilation to be made. Many field relations support the conclusion that the typical pluton (Fig. 56) represents a frozen and exhumed magma chamber which made space for itself by combinations of processes such as stoping (breaking off and sinking of fragments of wall rock, now preserved as xenoliths), assimilation (melting and/or dissolution of wall rock), or forceful intrusion (pushing aside of wall rocks, with concomitant development of intrusion-related folds and fractures) (see also Fig. 58). In addition, the pluton has usually thermally altered the wall rocks in a zone varying from a few metres to a few kilometres in width, forming a contact metamorphic aureole, which also provides valuable information on the conditions of intrusion and cooling.

An example of the type of evolution a pluton may represent is shown on the pressure (depth)–temperature diagram on Fig. 57. The starting point is some location in the upper mantle (1), where local conditions (such as pressure release on deep fractures or temperature rise due to mantle convection) cause partial melting of the peridotitic rocks (1–2), which, experiments show, release a magma of basaltic composition. The magma rises to some level in the lower crust, aided by its relatively low density, assimilating crustal material on the way (3). Trapped in a first magma chamber it starts to cool (3–4), and the first minerals, rich in iron and magnesium, crystallize. At a later stage, the remaining melts, now poor in Fe and Mg and rich in Si and alkalis, are affected by earth movements or rise diapirically (buoyancy due to low density) and collect at a higher level in a higher magma chamber formed in an old consolidated metamorphic complex (5). Slow cooling in this position produces the granite pluton as we see it today, eventually evolving as a solid mass within the upper crust and cooling to near-surface temperatures during subsequent uplift and denudation (6). In the presence of sufficient water, this period (5–6) will be marked by hydrothermal alteration of the magmatic minerals (see Fig. 67) and hydrothermal veining. Future evolution may involve a further rise in temperature around any part of the pluton used as a HLW repository (7).

A similar evolutionary pressure, temperature and time sequence can be envisaged for the metamorphic rocks into which the hypothetical pluton was intruded. This sequence will include at least two temperature peaks, an earlier one in which the original rocks were physically and chemically reconstituted

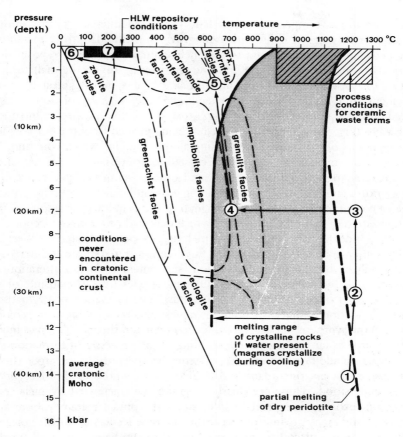

Fig. 57. Pressure (kbar) (depth)–temperature diagram with PT fields of the main groups of metamorphic rocks, the field of partial melting in the presence of water, and the fields representing the processing conditions of ceramic waste forms (e.g., Synroc, see Fig. 14) and the conditions at an early stage in the life of a HLW repository. Numbers 1–7 show the evolution of a hypothetical pluton, described in the text. (Data from Smith, 1982, and Best, 1982.)

during the period of intense folding (regional metamorphism, high-pressure, high-temperature mineral assemblages), and a later one in which the regionally metamorphosed rocks adjacent to the pluton are baked by the heat of intrusion (contact metamorphism, low-pressure, high-temperature mineral assemblages). The activity of fluids (H_2O, CO_2) in such metamorphic processes, which by definition take place in the solid state, below the melting temperatures of the rocks and minerals involved, can hardly be overemphasized and their availability largely determines whether the high-temperature minerals will revert to low-temperature equivalents (retrograde metamorphism) during subsequent episodes of uplift and cooling. Much of the data from metamorphic petrology is very pertinent to rock behaviour around HLW repositories, particularly in the low-pressure and low-temperature (hydrothermal) field.

The formation of the discordant dike and vein systems typical of every basement complex is difficult to generalize. Some veins are obviously associated with the pluton as radiating arrays (using radiating fractures as intrusion pathways) (Fig. 56) and are interpreted as being formed from the last volatile-rich remnants of the magma, containing all the elements which could not be built into the main minerals as they crystallized out. These veins are often extremely coarse-grained (pegmatites) and often contain valuable concentrations of ores and rare minerals (containing elements which were only present as trace constituents in the original magma). It is this ubiquitous feature of the crystallization of natural silicate melts which makes most geologists sceptical of the efficacy of the deep underground melting process (see Chapter 2, Section IV,D). Other mineral veins are due to fluids concentrated in the country rock and mobilized by the heat of intrusion, still others to *in situ* alteration of country rock by chemically active diffusing species (metasomatism), or to partial melting of the country rock (anatexis). The formation of pegmatites, mineral veins and primary ore bodies is typical of certain problems in geology in which the synthesis of observation, experiment and theory still leaves much to be desired as far as a deep understanding of the processes of formation are concerned. In addition to these vein systems, however, there are larger, sheet-like, discordant or concordant bodies (dikes, sills) whose mode of origin is better understood. These are clearly magmatic (characterized by melt crystallization textures, fine-grained cooled margins, contact metamorphism and xenoliths) and many are interpreted as the one-time feeders of surface volcanic eruptions. Some, of course, may not belong to the basement at all, in the sense that they are so young that they also cut through the cover sediments: an example would be the Tertiary dike which intersects the bedded salt formations in the vicinity of the WIPP site (see Fig. 50).

B. Example: Götemar Granite, Sweden

Granites are very common crystalline rocks consisting mainly of quartz and felspars; they often occur in bodies (plutons) of large size and simple, homogeneous internal structure. Granite is thus one of the main host rocks under consideration for HLW repositories in many countries (Canada, Sweden, Switzerland, the United States), with detailed work in progress on many fronts (see Chapter 2, Section IV,B). As an example we take the Götemar granite near Oskarshamn in south eastern Sweden, part of which has been studied in detail as a possible site for a deep-mined HLW repository (Kråkemåla site) (see Fig. 94). The coarse-grained variety has been quarried and exported around the world as a decorative, monolithic building stone for over 50 years. Its favourable qualities, which include uniformity (grain size ~ 1.5 cm), lack of alteration, and sparsity of joints (mean spacing 1–5 m), together with the excellent exposure (60% of the area consists of exposed bedrock surfaces), make the site particularly attractive as a prospective repository location.

The Götemar pluton is almost circular in section (Fig. 58) and was emplaced in its present surroundings, consisting of older, already crystallized granites of different types (Småland granite suite), about 1380 m.y. ago. The coarse-grained variety of the granite makes up most of the pluton and consists of 20–35% quartz and 60–75% felspar (intergrowths of microcline and albite in roughly equal amounts), with 5% or less of other minerals (mainly biotite, with accessory minerals such as fluorite, zircon, and apatite). Chemical analyses indicate conditions of crystallization of 2 kbar (7–8 km depth) at just over 700°C, with HF playing an essential role among the volatile constituents. The other varieties (medium-grained, fine-grained, and porphyritic) show irregular distributions within the pluton. They seem to represent later stage differentiates of the same magma and have suffered most from late-stage hydrothermal alteration. Highly mobile fluids remained when crystallization was almost complete and invaded the pluton and its surroundings, forming concentrically and radially arranged granite and pegmatite veins which contain concentrations of rare minerals (fluorite, topaz, and beryl) and common ores (pyrite and galena).

East of the major fault which later intersected the pluton (Fig. 58), the contact between the Götemar granite and its surroundings is generally sharp and well defined. There is no sign of marginal chilling (decreasing grain size towards the contact), and there are few structures (xenoliths or flow banding) which can be related to the process of intrusion. However, a well-developed pattern of primary joints is developed (fracture planes with coatings of the same minerals found in the pegmatites) with concentric, radial or subhorizontal preferred orientations (Fig. 58). These joints, like the related shear zones and pegmatite veins, also affect the rocks surrounding the granite, becoming less common and more widely spaced outwards from the contact. All these relationships suggest crystallization at a lower level and subsequent intrusion as a crystal mush with a high volatile content at a subvolcanic level. To the west of the major fault, the presently exposed part of the pluton represents the deeper levels of the same body, as evidenced by the felspathization of the wall rock (metasomatism), the more intensive pegmatite veining (extending up to 300 m from the contact), the more irregular, ill-defined form

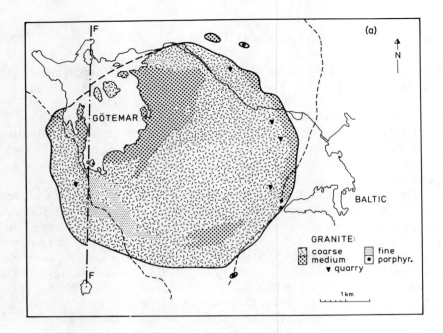

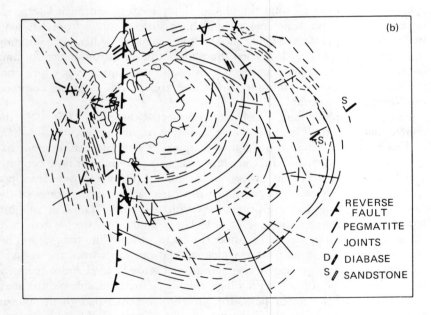

Fig. 58. Outline maps of the Götemar pluton on the southeastern coast of Sweden, showing the distribution of (a) the different granite types and (b) the main joint zones and pegmatites. One of the sites being studied in Sweden as potentially suitable for a HLW respository is located towards the eastern margin of the pluton (Kråkemåla, see Fig. 94). (From Kresten and Chyssler, 1982, by permission of the Geological Society of Sweden.)

of the contact itself, and the lack of a well-developed primary joint pattern. The vertical displacement on the fault is probably at least 500 m, but a related system of secondary (postintrusion) joints seems to be absent.

IV. CERAMIC WASTE FORMS AND MINERAL ANALOGUES

Ceramic waste forms are essentially synthetic crystalline rocks in which the radionuclides are bound in the crystal lattices of the various mineral-like phases. They are manufactured by crystallizing melts or recrystallizing powdered aggregates at high temperature and have proved to have a higher short-term stability than glass under experimental hydrothermal conditions (see Chapter 8, Section IV). Also, if the mineral-like phases are arranged to be chemically and/or structurally similar to minerals which occur in nature in a wide range of geological situations, ceramic waste forms would potentially provide much greater confidence in very long-term stability. In addition, it has recently been demonstrated that ceramics allow a higher waste loading (waste solids 50–90% by weight) and simpler processing (usually no molten materials and processing in sealed containers) than glasses, which would tend to reduce overall waste management costs. However, research and development on ceramic waste forms is at a relatively early stage, and the momentum of the vitrification program has dictated that the first generation HLW monoliths will be glass.

Ceramic waste forms have been produced in the laboratory and in small processing–research facilities in a wide range of compositions and by a variety of processes. In general, however, two steps are envisaged. First, the wastes are chemically processed, solidified and compositionally adjusted by various additives to give a starting material of the required composition, generally in the form of a dry powder or a glass. Second, the aggregate is subjected to heat treatment, sometimes under high pressure, to produce a dense, crystalline end product. At one end of the scale we have the direct melting of incombustible solid radwastes in an electrofurnace at temperatures of 1300–1500°C and subsequent cooling to a synthetic igneous rock (e.g., synthetic basalt produced at the Idaho National Engineering Laboratory from transuranic-contaminated concrete, metal, sludge, and soil). At the other end, there are the processes at elevated temperatures (150–250°C) and pressures which use cement or concrete as a matrix for radwaste solids (e.g., the FUETAP concretes produced at the Oak Ridge National Laboratory). Between these two extremes, we find mixtures of waste solids and inert additives being converted to crystalline aggregates by such processes as Pyroceram (melting, glass formation, and subsequent controlled devitrification), fusion-casting (melting and controlled crystallization), sintering (firing to temperatures of 900–1200°C under atmospheric pressure) and hot-pressing (firing under high hydrostatic or nonhydrostatic pressure). Examples of two of these processes are given in Figs. 14 and 59. The ultimate aim is to produce a ceramic in which all the radionuclides are incorporated in stable crystal structures and in which all the phases have stable mineral analogues whose behaviour is known in nature.

The main mineral analogues which have been considered up to now are listed in Table IV. Of the 2500 mineral species known, only ∼100 come into question as hosts from the point of view of insolubility in aqueous solutions, suitability of crystal chemistry, and simplicity of chemical composition. Of these, few are of common enough occurrence that their long-term stability can be assessed in a variety of natural situations and even fewer are stable at low temperatures in the presence of groundwater (i.e., under repository-like conditions). Most of the minerals listed are associated with pegmatites and alkaline magmatic rocks (i.e., formed from magmas impoverished in silica after repeated fractional crystallization) and, perhaps surprisingly, few are silicates. Silicate structures utilize the most abundant elements of the Earth, but the critical radionuclides, with the exception of ^{90}Sr, are unusual elements whose nuclei are either too large or too small to fit into available sites in the silicate minerals. The association of suitable mineral analogues with pegmatites is thus to be expected since, as pointed out above, they form from the

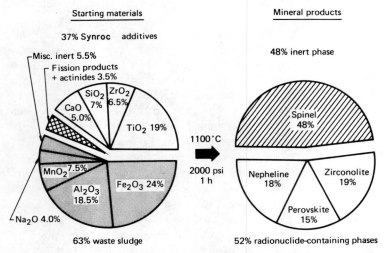

Fig. 59. Chart showing the wt% composition of the starting materials and the synthetic crystalline phases in SYNROC-D, proposed for the immobilization of defense wastes from the Savannah River Plant. (From Campbell *et al.*, in Lutze, 1982, by permission of Elsevier Science Publishing Co.)

Table IV. List of the Main Mineral Analogues Being Considered as Host Phases for Various Radionuclides in Ceramic Waste Forms[a]

Mineral or structure type	Illustrative compositions	Radionuclides incorp.	Occurence of mineral in nature	Stability
Apatite	$Ca_5(PO_4)_3(OH,F,Cl)$ $(Ca,Sr)_5(PO_4)_3$	Sr, Ln, (Ru)	Accessory min. in igneous/metamorphic rocks, phosphate deposits	Very stable (detrital mineral in sedimentary rocks)
Baddeleyite Cerianite	ZrO_2 $(Ce,Th)O_2$ $(Ce,Zr)O_2$	Ln, Act	Alkaline igneous rocks and carbonatites, pegmatites, kimberlites	Stable (not metamict, formed during weathering and hydrothermal alteration)
Hollandite	$BaMn_7O_{16}$ $BaAl_2Ti_6O_{16}$ $(K,Ba)(Ti,Fe)_8O_{16}$	Cs	Manganese ores	(Data insufficient)
Monazite	$(Ce,La)PO_4$ $(Ln,Act)PO_4$ $ActSiO_4$	Ln, Act	Pegmatites and hydrothermal veins, accessory min. in granitic rocks	Very stable (not metamict, placer deposits, detrital minera in sedimentary rocks)
Nepheline	$(Na,K)(AlSiO_4)$	(Cs)	Alkaline volcanic and plutonic rocks as main constituent (e.g., neph. syenite)	Stable

(*Continued*)

Table IV. (*Continued*)

Mineral or structure type	Illustrative compositions	Radionuclides incorp.	Occurence of mineral in nature	Stability
Perovskite	$CaTiO_3$ $(Sr,Ba)(Ti,Zr)O_3$	Sr, Ru, Tc^{4+}, Ln, (Cs, Act)	Alkaline and basic igneous rocks, carbonatites	Relatively unstable (metamict, easily altered, but still retains Ln)
Pollucite	$CsAlSi_2O_6$ $(Cs,Na)AlSi_2O_6 \bullet nH_2O$	Cs	Pegmatites	Relatively unstable (not metamict, but alters to clay minerals by weathering)
Scheelite	$(Sr,Ba)MoO_4$ $CaWO_4$	Sr, Tc^{6+}, (Ln, Act, Ru, Te)	Pegmatites, contact metamorphic rocks	(Data insufficient)
Spinel	$MgAl_2O_4$ AB_2O_4 where A = Mg,Fe,Ni,Co B = Al,Cr,Ti	Tc^{4+} (Ru) process contam.	High T metamorphic rocks, accessory min. in gabbros, peridotites	Very stable (not metamict, detrital mineral in sedim. rocks)
Titanite (sphene)	$CaTiSiO_5$	Ln, Sr, Cs	Accessory mineral in granitic rocks	Very stable (not metamict, residue in weathered zones, concordant U/Pb ages, etc.)
Zirconolite	$CaZrTi_2O_7$	Sr, Ru, Ln, Act	Gabbros and pyroxenites, carbonatites, kimberlites	Relatively stable (metamict, often altered, but retains Act, see Fig. 60)

[a] Main sources: Roy, 1982; Haggerty, 1983; Grauer, 1984.

residual fluids remaining after the crystallization of the main silicates from the magma and thus contain all the species which did not "fit." Although these minerals formed at temperatures of up to 600°C and pressures up to several kilobars in nature, it was not the pressure–temperature regime that was critical to their formation, but rather the complex chemistry of the solutions.

A particularly important aspect of waste form technology is radiation damage. This has been extensively investigated in both vitreous and ceramic waste forms by the technique of "doping," i.e., subjecting the material to the equivalent of long-term dosages by incorporating short-lived, highly radioactive substitutes for the long-lived waste radioisotopes (e.g., ^{238}Pu for ^{239}Pu). Many different radiation effects have been studied quantitatively (including changes in density, stored energy, and microstructure, microfracturing and helium buildup) but no significant increase in experimental leach rate has been observed. Some minerals, however, show a transformation from the crystalline to an amorphous state, called metamictization, which is thought to be due to radiation damage, and there is evidence that the metamict forms of some minerals are more susceptible to hydrothermal alteration than their crystalline equivalents. In the following, we focus on this aspect of mineral analogue studies as an example of the direct application of geological data to long-term stability assessment.

Example: Zirconolite

Synthetic zirconolite, $CaZrTi_2O_7$, is one of the main constituents of the ceramic waste form Synroc (see Fig. 59), in which it serves as a host for actinides and, to a lesser extent, for Sr and lanthanides. In nature, it occurs as a member of the pyrochlore series in rocks derived from the late-stage differentiates of basic magma (alkali syenites, alkali granites and carbonatites) and in kimberlite pipes. Natural zirconolite can contain up to 23% Th, up to 10% U, and up to 13% Y plus lanthanides (wt % oxide) and occurs in both metamict and crystalline states. Although the process of metamictization is not completely understood, it seems to be mainly due to lattice damage caused by α-particle emission by, and recoil of, the long-lived radionuclides bound into the crystal lattice. The amount of lattice damage is thus related to the total number of α-emissions since the mineral crystallized, a parameter which can be calculated from a knowledge of the radioisotope contents, their decay constants, and the time since crystallization.

Figure 60 shows the α-dose–time curve for zirconolite in Synroc (zirconolite plus perovskite plus hollandite) containing 10% waste solids in the form of HLW calcine from a reprocessing plant. Analysis of the U and Th contents of natural zirconolites (giving the total α-dose) and of the U and Pb isotopic

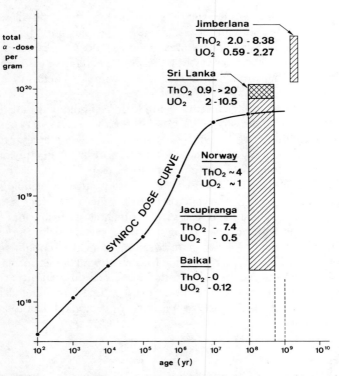

Fig. 60. Total α dose versus age for natural zirconolites and perovskites and the dose curve for proposed synthetic equivalents in Synroc ceramic waste forms containing 10% HLW solids. (Data from Oversby and Ringwood, 1981.)

compositions (giving the age) shows that the least lattice damage is equivalent to a Synroc age of 10,000 years. Many zirconolites have suffered radiation damage far in excess of anything a synthetic equivalent would be subjected to in Synroc. Of particular interest are the Sri Lanka samples, which have extremely high U and Th contents and are completely metamict. Different specimens and different isotope pairs $^{206}Pb/^{238}U$, $^{207}Pb/^{235}U$ and $^{207}Pb/^{206}Pb$) give closely similar isotopic ages, indicating that they have remained as closed systems to U, Th and Pb for 550 m.y. (Fig. 60). In general, data from natural zirconolites indicate that migration of U, Th, and Pb has been negligible over times in excess of 10^8 years and in a variety of geological situations which include elevated temperatures and pressures, contact with groundwater, and weathering and erosion, and that is true whether they are metamict or not.

V. CONCLUDING REMARKS

In this chapter we have surveyed naturally occurring crystalline rocks and synthetic equivalents used in radwaste disposal technology, from the point of view of the conditions under which they form. We have been dealing with temperatures of several hundreds to over a thousand degrees Celsius and pressures of up to several kilobars. This corresponds to depths in the Earth's crust of many kilometres, out of range of direct observation, only amenable to deductions based on the structure, texture, mineralogy and chemistry of frozen and exhumed representatives of past activity. The enormous body of data which now exists from the magmatic and metamorphic complexes of the world, and the resulting understanding of deep crustal processes, may make a significant contribution to prediction of the long-term behaviour of heat- and radiation-producing wastes in deep repositories of any kind, whether in crystalline rocks themselves or in not yet metamorphosed sedimentary equivalents. In all situations, the role of fluids is often enigmatic, but clearly of utmost importance. This will be a recurring theme in the following chapters dealing with present-day processes in the upper part of the Earth's crust.

SELECTED LITERATURE

Formation of crystalline rocks (magmatism, metamorphism, metasomatism): Bailey and MacDonald, 1976; Wyllie, 1980; Best, 1982; Smith, 1982 (pp. 51–92); Granath *et al.*, 1983.

Granite as host rock for test facilities and HLW repositories: Kresten and Chyssler, 1976; Swedish Nuclear Fuel Supply Co., 1978; Smedes, 1980; Witherspoon *et al.*, 1981; Gonzales, 1982; Lawrence Livermore Laboratory, 1982; Heuze, 1983; Organization for Economic Cooperation and Development (OECD), 1983c.

Long-term behavior of ceramic waste forms and mineral/rock analogues (including radiation effects): Ewing, 1976; McCarthy, 1979 (pp. 169–236); Northrup, 1980 (pp. 239–322); Moore, 1981 (pp. 123–200); Oversby and Ringwood, 1981; Ringwood *et al.*, 1979; Johnston and Palmer, 1982; Lutze, 1982 (pp. 299–418); Roy, 1982 (pp. 36–112, 184–232); Topp, 1982 (pp. 641–732); Grantham *et al.*, 1983; Haggerty, 1983; Kesson and Ringwood, 1983; Weber and Roberts, 1983; Grauer, 1984; Dole *et al.*, 1983.

CHAPTER 10

Physical Processes in the Upper Crust

I. INTRODUCTION

The last three chapters (7–9) have been mainly concerned with the way in which different types of rock have formed in the past and with the relevance of these rock-forming processes to different aspects of the radwaste disposal problem (rock as enclosing medium, natural analogues, waste form technology, potential hazards, etc.). We now return to the Earth's crust as it exists today. In two previous chapters (5 and 6), we discussed the chemical and physical processes going on today at the Earth's surface and within the regolith, and their long-term effect on surface repositories. Now we are in a position to penetrate below the surface layers into the bedrock and to consider the environment into which a deep underground repository would be emplaced and the perturbations such a repository would cause. The purpose of this chapter is to discuss the physical aspects of the upper few kilometres of the bedrock environment. The chemical aspects are treated in Chapter 11.

The line of argument in this chapter is as follows. First, we look at the main physical parameters (temperature, pressure, and stress) and their variations within the upper part of the crust. Some of the complicated interrelationships between these parameters and the properties of rock masses (such as thermal conductivity, strength and porosity) are then pointed out, before considering the most important site-specific environmental factors, structural state and fluid flow field. By structural state is meant the cumulative result of past crustal movement as evidenced by the present structural pattern (faults, joint systems, and fold and diapir structures). Fracture analysis is a particularly important aspect of the repository site characterization and, since fracture formation causes earthquakes, is closely related to the question of seismic hazards. Under fluid flow field is included the pattern of deep groundwater circulation, either as a topographically controlled meteoric system or a convective system driven by a deep heat source. It becomes clear that the structural state and fluid flow field of a particular crustal segment are intimately connected and are both affected by the construction of a deep cavity and the emplacement of heat-producing radwaste. The complex thermomechanical, hydromechanical, and thermohydrological effects of a deep HLW repository are thus the subjects of the final section. The approach is not that of a systematic treatment but rather a presentation of selected examples within a connecting framework.

II. PHYSICAL PARAMETERS

The main parameters affecting rock behaviour at depth are temperature, lithostatic pressure, pore fluid pressure, and stress. Transport of heat, material, and fluids depends on gradients in any of these in space and variations with time. It is clear that wide ranges in these parameters occur naturally in the crust (e.g., local very high temperatures near volcanic feeders), but there are certain values which can be considered broadly average or typical of a certain geotectonic situation over large areas and for specific depths in the crust. These typical values and ranges are now well known from measurements in deep mines, from borehole logs, and from remote sensing using the enormous variety of techniques and tools developed over the past 100 years for subsurface mineral, energy, and water resource exploration (see Chapter 14, Section II).

Crustal temperatures may be estimated from mine and borehole measurements, heat flow observations, thermal conductivity determinations on rocks and minerals, and indirect petrological and geochemical evidence. Near-surface geothermal gradients in stable cratonic areas average $\sim 30°C/1000$ m, but vary from $\sim 20°C/1000$ m in some rapidly subsiding sedimentary basins to $> 50°C/1000$ m in geothermal areas associated with magmatism and hot circulating fluids at relatively shallow depths. The ambient temperature for most deep-mined repositories will thus generally be $< 50°C$. In low-relief cratonic areas, the isothermal surfaces are subhorizontal and heat flow takes place mainly by upward conduction (in contrast to transport with moving fluids, as in volcanic and geothermal areas). The thermal conductivities of rocks are generally low, mainly in the range of 2–3 W/m • sec • deg, but can show wider ranges with complicated variations depending on mineralogy, temperature, or moisture content. Salt, for example, has an unusually high conductivity at low temperatures (5.5 W/m • sec • deg at 25°C), decreasing to normal values at high temperatures (e.g., ~ 3 W/m • sec • deg at 250°C). In this context, it is interesting to note that glass has a much lower thermal conductivity than most rocks (<1 W/m • sec • deg at temperatures $< 100°$C). Isothermal surfaces in the upper crust are not precisely horizontal (i.e., heat flow varies gradually from place to place). This is due to the natural distribution of radioactive isotopes in the Earth's crust (particularly ^{238}U, ^{235}U, ^{232}Th, and ^{40}K) and their concentration as trace elements in certain rock types (particularly in acid and intermediate magmatic rocks), as well as to many other types of heterogeneity.

The pressure due to the overlying rock column (lithostatic pressure) at some point in the crust can be calculated theoretically from the densities and thicknesses of the rock formations and from the depth. The average density of upper crustal rocks is 2.67 g/cm^3, and many of the most common rock types (granites, limestones and sandstones) approximate to this value, although some rocks are significantly heavier (e.g., basalt, 2.9–3.0 g/cm^3) or significantly lighter (e.g., salt, 2.1–2.2 g/cm^3). From this it can be estimated that the lithostatic pressure gradient in granitic basement should be ~ 260 bar/1000 m, whereas in thick sedimentary sequences it would be slightly less and probably nonlinear. However, it is doubtful whether such theoretical values are meaningful in the upper few kilometres of the crust, since *in situ* measurements show large deviations and indicate that the basic assumption that rock can be regarded as a very viscous fluid is false, if not viewed over very long periods of time. Of greater short-term significance is the pore fluid pressure, which is known from direct borehole determinations to vary between hydrostatic (weight of water column) and lithostatic (weight of rock column). Between these limits, the pore pressure at depth depends strongly on local conditions of fluid density and viscosity, porosity and permeability, compaction rate, and water production by dehydration and other mineral reactions. Very high pore pressures are usually associated with high-porosity and low-permeability rocks (e.g., shales), whereas highly permeable rocks (jointed granites and limestones or porous sandstones) generally show sufficiently interconnected pathways to the surface that the pore pressure remains low (i.e., approximately hydrostatic). In crystalline rocks, permeability decreases systematically with depth in the upper few hundred metres, but below this it varies unsystematically (hydraulic conductivities between 10^{-4} and 10^{-6} cm/sec) and is still high compared with shale (10^{-9}–10^{-11} cm/sec).

This means that pore pressures greater than hydrostatic are unlikely at any depth in terrains where crystalline rocks outcrop.

Most rock volumes in the upper crust are in a deviatoric stress field, i.e., subjected to different pressures in different directions. Maximum principle stress differences can be inferred from various lines of evidence and measurement techniques and suggest maximum values of 2000 bar. Hence, the deviatoric stress will often be far in excess of the nominal lithostatic pressure at the depth of most proposed repositories. Complex relations exist between stress buildup, depth (temperature, confining pressure), permeability, and pore fluid pressure, and these determine the mechanical behaviour of rocks (strength and failure characteristics) in natural and in artificial (i.e., repository) situations. These relations have been studied intensively on small, intact specimens in the laboratory (Fig. 61) but are poorly known for large rock volumes with their many discontinuities.

III. ROCK DEFORMATION AND STRUCTURAL STATE

The various physical parameters above are intimately connected and subject to change with time. The temperature and pressure conditions determine, together with the mechanical properties of the various rock types, the reaction of the rock complex to the deviatoric stress. If the stress difference exceeds a certain amount (the strength of the rock), deformation takes place, either by fracture (brittle deformation) or by flowage (ductile deformation). This is strongly influenced by pore fluid pressures (Fig. 61): high pore pressures promote brittle failure and movement on already formed fractures and lower the strength of the rock. Once deformation takes place, further changes in fluid transport regime and in temperature and stress fields are initiated. Although deep repositories will be sited in tectonically stable areas, i.e., areas in which regional stresses have long been far below the bulk strength of the crust and are likely to remain so in future, two types of past deformational event are important because of the resultant structural state. First, "hard" rocks (crystalline rocks, welded tuffs or basalts) tend to show intricate systems of fractures due to past (and sometimes continuing) brittle deformation, fracture systems which control the present-day deep groundwater flow pattern. Second, "soft" rocks (shale and salt) tend to show complicated flow structures due to past ductile deformation, of which the most important from the point of view of radwaste are the salt diapirs which are being considered as repository hosts in some countries. We focus here on these particular aspects of crustal rock deformation.

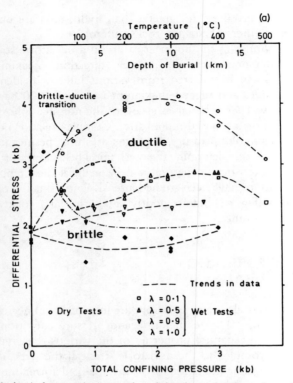

Fig. 61. Mechanical properties (strength and ductility) of rocks as determined in the laboratory by testing small, intact rock specimens. (a) Results of constant strain rate (3×10^{-5}/ sec) tests at various temperatures and pressures (corresponding to a geothermal gradient of 30°C/km) and at various pore pressure/confining pressure ratios (λ). Increasing λ-values reduce the rock strength and cause a change from ductile to brittle behaviour at constant depth. (b) Strength of various types of rock as a function of temperature at geologically realistic strain rates for crustal movements, based on experimentally determined flow laws for polycrystalline aggregates. (From Rutter, by permission of Elsevier Publishing Co.; Schmid, in Hsü, 1982.)

A. Brittle Deformation: Fault and Joint Systems

At shallow crustal levels, most rocks lie within the field of elastic–brittle behaviour in response to natural stress buildup mechanisms in intraplate environments. For this reason, particularly "hard" rocks are typically intersected by systems of fractures of various types and sizes, due to processes of flexure, uplift, cooling, extension and compression, and these control the fluid transport properties of the rock mass. Even rocks generally considered to show long-term plasticity (shale and salt) and, in places, unconsolidated materials

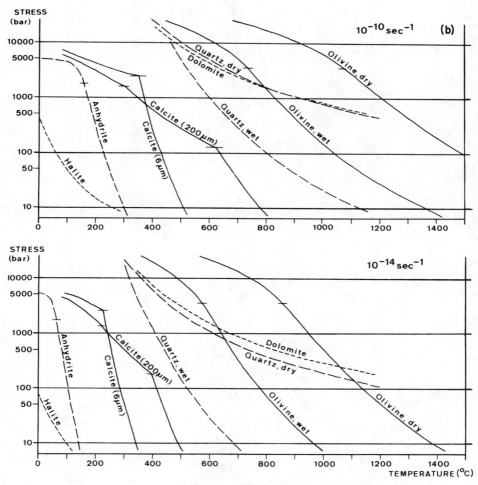

Fig. 61. Continued.

(glacial till) sustain systems of cracks for sufficiently long times to affect significantly their hydraulic behaviour. The spatial variability of the subsurface fracture systems is one of the more troublesome features of any potential repository site (see Chapter 14, Section II).

Two fundamentally different types of fracture develop under the confining pressures typical of shallow depths: shear fractures and tension fractures. On a shear fracture, relative movement is lateral along the fracture plane and is often indicated by displacements of lithological features in the wall rocks (e.g., displacement of veins) (see Fig. 62) and by striations in the direction of move-

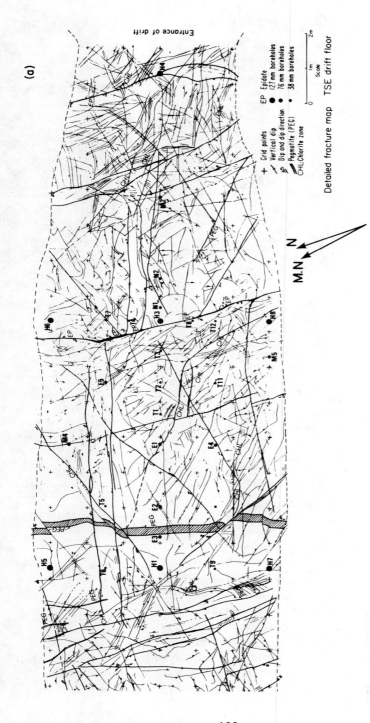

Fig. 62. Jointing and fracturing of granitic rocks in the Stripa Test Facility, Sweden. (a) Detailed map of discontinuities in the floor of the time-scale experiment drift. (b) Vertical section containing data from the boreholes in the floor of the time-scale experiment drift. (From Thorpe, 1979.)

192

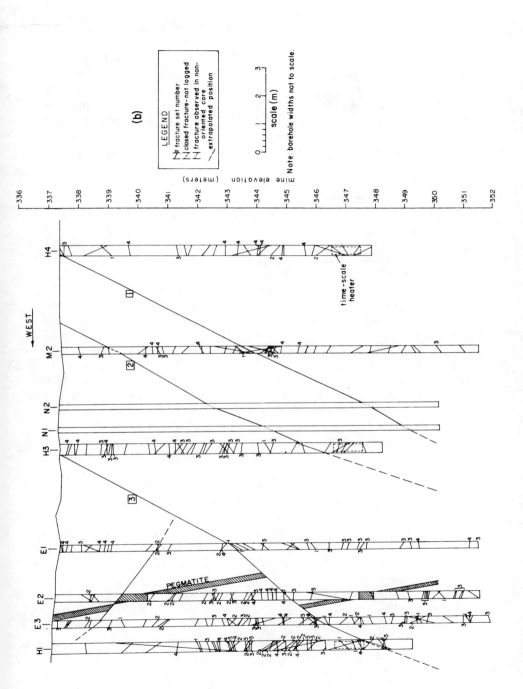

ment (polished surfaces with scratches and grooves or fibrous mineral growth). In contrast, tension fractures show no lateral displacements or striations because the fracture walls move apart, leaving a void which is subsequently partially or completely filled with minerals. In practice, however, this fundamental distinction is often obscured (movement on irregular shear fractures produces voids, tension cracks form planes of weakness for later shearing, etc.), and brittle reaction to a particular stress field results in an intricate system of discontinuities showing a certain geometric regularity but defying a strict mechanical classification. From the geometry of a fracture system, some characteristics of the original stress field can be deduced (e.g., the orientation of the principal stress axes). Similarly, extensive laboratory testing, computer modelling, and theoretical analysis have enabled the prediction of some aspects of potential fracture patterns in new stress fields, for instance, around deep-mined cavities and tunnels.

In geology, fractures are roughly classified according to size. Faults are large fractures, generally visible as lines of discontinuity on geological maps and as zones of crushed rock (fault gouge) in the field or in underground workings. In intraplate situations, the stress field results in two main types, both subvertical near the Earth's surface. Normal or dip–slip faults form by subvertical relative displacement of the fracture walls, whereas transcurrent or strike–slip faults are due to lateral, subhorizontal movement. However, many large faults show long, complicated movement histories, with both horizontal and vertical displacement components. Joints are small fractures, visible as lines of discontinuity or as very thin mineral veins in outcrops, cores, and underground workings (Fig. 62). They are often curved or irregular and may show little or no displacement of intersected discontinuities or lithological boundaries. Most joints are at least partially lined with hydrothermal or even lower temperature minerals, indicating extensive, fluid-assisted chemical migration subsequent to joint formation (see Chapter 8, Section IV,B). Faults and joints which are temporally related show compatible geometries and many transitions (minor faults, shear zones, or master joints), and are accompanied by microscopic or submicroscopic fractures, generally referred to as cracks or microcracks. The faults, joints and cracks formed during one deformational event build a three-dimensional network of discontinuities with considerable variations in space which are practically impossible to predict. This is best illustrated by observing the fracture terminations.

Fractures of a particular size have finite lateral extents (Fig. 62) and show various types of termination, of which the most common are transformational (shear fracture transformed to a differently orientated tension fracture and *vice versa*), dispersive (breakup into successively smaller and more numerous fractures of the same type), coalescent (convergence with a larger fracture of the same type) and transitional to ductile deformation (change to deformation by crystal plasticity and diffusional flow). The resulting variations in spacing and size of fractures is complex even within a homogeneous rock body, such as a granite pluton, and becomes even more so in a heterogeneous body, with different rock units showing different mechanical properties. This complexity in space is also a reflection of the complex evolution of brittle fracture with time known from experimental rock deformation. Cylindrical specimens subjected to a steadily increasing axial compression under low-pressure and -temperature conditions go through various stages: (1) closing of pre-existing cracks and collapse of pores; (2) elastic deformation of grains, possibly some slip on pre-existing cracks; (3) proliferation and stable propagation of increasing numbers of newly formed microcracks; and (4) unstable microcrack propagation, local weakening, linking of microcracks, and localization of macroscopic fractures. The transition from 3 to 4 is accompanied by dilatancy (volume and porosity increase due to forced movement on irregular cracks) and acoustic emission (microseismicity). The whole process is facilitated (i.e., takes place at lower stress differences) if the specimen contains pore fluids under high pressure. Increasing the pore fluid pressure lowers the rock strength and promotes brittle as opposed to ductile failure (Fig. 61), a fact which is of enormous importance for geology (earthquake prediction, mechanics of faulting) and for subsurface activities of all types. This leads to a short digression into the relation between fracture formation and seismic activity, as applied to the safety of underground repositories.

B. Brittle Deformation: Seismic Hazard

Brittle deformation in the Earth's crust is the cause of seismic activity on all scales, from the major earthquakes associated with large-scale fault zones (plate boundary systems) down to microseismic events only discernible with the most sensitive instruments. Present-day or historical seismicity is an indication of the formation of new fractures or the reactivation of old fractures at depth, and the hazard potential for an underground repository is a function of the energy released (the magnitude of the earthquake), the distance from the point of release (the epicentre of the earthquake), and the vibration propagation characteristics of the intervening rocks. The first rule in site selection is therefore to avoid active seismic zones, i.e., to avoid plate boundary systems. Nevertheless, even in aseismic areas, large earthquakes are known to occur, sometimes but not always related to known intraplate fault zones, and attempts are being made to determine the probability of occurrence of an earthquake of a certain size in such areas and to assess its possible effects in an underground repository.

Figure 63 is an earthquake risk map of Switzerland, showing the intensity on the MSK scale of the largest earthquake to be expected in the next 10,000 years. The region at present under investigation for possible deep-mined repository sites was chosen partly because of its low seismic risk as indicated by this map (see Chapter 15, Section V). However, maps of this type in areas of low seismicity are based on numerous assumptions of questionable validity, including extrapolation of poor historical records, identification of "seismic areas" on poor geological evidence of active faulting and estimation of an upper limit for a possible earthquake magnitude, and should not be read too literally. In most areas, a major seismic event is likely to occur within the lifetime of any HLW repository and some attention must be paid to possible effects. It is now generally agreed that the damaging effects of the ground vibrations themselves decrease with depth, and experience in mines suggests that they will be unimportant, except possibly during the emplacement phase and under special circumstances. Of more concern are the indirect effects related to the associated movement on one or several faults and to the development of new systems of cracks and joints, which in turn strongly affect the hydrogeological regime (dilatancy pumping, new hydraulic pathways and abnormal pore fluid pressures). Further indirect effects also affecting the groundwater system are due to the surface changes caused by large earthquakes, such as landslides causing flooding or changes in drainage systems or tsunamis. In general, the effect of seismicity on the long-term stability of a deep repository is one more aspect of the mutual interaction between brittle deformation and fluid flow, between crustal movements and the underground plumbing system, which is important in many other areas of earth science.

C. Ductile Deformation: Salt Diapirism

The flow of rock salt into domal structures or diapirs is an example of ductile rock deformation and produces structures similar to those typical of the cores of orogenic belts (ductile deformation of silicate and carbonate rocks at convergent plate boundaries) and in glaciers and ice sheets (ductile deformation of ice). In all these situations, the rock deforms by solid-state creep, changes in shape being accommodated by molecular diffusional processes, migration of dislocations, twinning or pressure solution under conditions of temperature, confining pressure, pore fluid pressure and stress which do not allow purely mechanical fracture and slip. Differences in the mechanical properties of different rock layers (often expressed as different "viscosities"), original irregularities in the shapes of the layers, and spatial and temporal variations in the stress field lead to the development of typical deformational structures. Intense folding develops when the cumulative result of the stresses was one of compression parallel to the layering. In general, this results in folds with almost rectilinear hinges (fold axes) and almost planar limbs, arranged in a

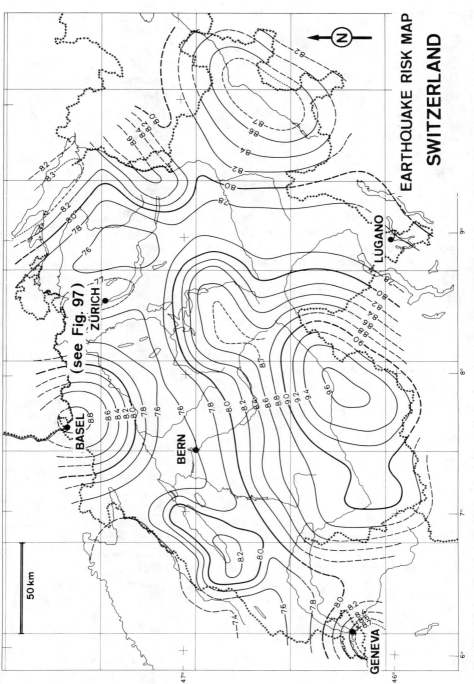

Fig. 63. Earthquake risk map of Switzerland showing lines of equal intensity (on the MSK scale) for a probability of occurrence of 10^{-4}/year (one earthquake of the given intensity every 10,000 years). The Basel earthquake of 1356, the largest historical one in Europe north of the Alps, had an estimated intensity of 9–10. (By permission of Institut für Geophysik, ETH, Zurich.)

series of waves with wavelength, amplitude, and symmetry determined by local factors such as layer thickness, viscosity contrasts and stress orientation. If the cumulative effect of deformation was one of extension parallel to the layering, differential necking and pulling apart of the layers takes place (boudinage). Even in the absence of compositional layering, rocks take on a penetrative fabric due to preferred orientations of individual mineral grains. Thus, most deformed rocks show a marked foliation, cleavage, or plane of easy splitting, which is also the plane of flattening of objects of known original shape (fossils, pebbles, xenoliths, etc.). Some deformed rocks also show a lineation due to the preferred orientation of prismatic mineral grains and a concomitant stretching of deformed enclosed objects. Folds, boudins, foliations and lineations often occur together in complex geometrical arrays which can be analysed, like the arrays of faults and joints mentioned above, to yield information on the deformational histories and mechanisms of the rocks involved. For salt domes and plugs, analysis of their complex internal structures together with model experimentation, laboratory testing, theoretical analysis, and field investigation of other aspects (age dating or host rock stratigraphy) has led to a consistent picture of formation by a process known as diapirism.

The driving force of diapirism is gravity working through the buoyancy of materials which are less dense than the ones overlying them. The effects of such buoyant forces can be seen in many different geological situations (e.g., magma rise and intrusion, gneiss dome formation, or mantle convection), but they are best known from their effects on salt beds. Salt becomes less dense than other sediments below depths of about 1 km (due to the increase in density of the other sediments during compaction, dewatering and diagenesis) and becomes much weaker and more ductile than other sedimentary rocks at depths of 2 km or more. Field data indicate that when a salt bed is thick enough (300–400 m) and when it becomes sufficiently deeply buried ($>$ 1–3 km depending on local conditions), lateral flow takes place in the salt bed under the influence of original thickness variations, overburden pressure variations and/or tectonic stresses. Salt ridges and mounds develop, leaving broad troughs between them and causing related changes in the sedimentary environment at the surface. Whether the mounds are allowed to grow into domes depends on the mechanical properties of the overlying beds and on other local factors (such as faulting, fluid content and density differences). Under favourable circumstances, a dome eventually pierces the overlying layers and moves upwards as a plug or salt diapir (Figs. 64 and 65). The pierced units sag into the vacated space, giving typical peripheral sinks around the diapir and tensional faulting across the roof. The rate of upward movement of salt diapirs during the piercement phase is estimated at 1–20 m/1000 years when averaged over long periods of time (strain rates of 10^{-11}–10^{-15}/sec, at stress differences of 5–15 bar). However, the very complicated

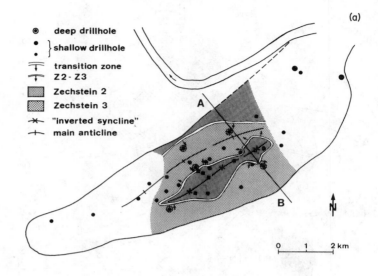

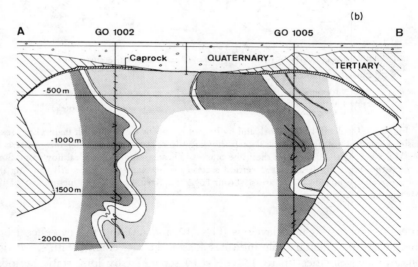

Fig. 64. Internal structure of a salt diapir as determined by exploratory drilling, Gorleben, West Germany. (a) Map of the top of the diapir, based on numerous shallow boreholes through the overlying Tertiary and Quaternary deposits. Clearly visible is the inverted anticline, an upfold with younger rocks (Zechstein 3) in the core. (b) Reconstructed section through the diapir based on geophysical surveying and two deep boreholes, showing the complex internal structure. (From Bornemann, 1982, by permission of Deutsche Geologische Gesellschaft.)

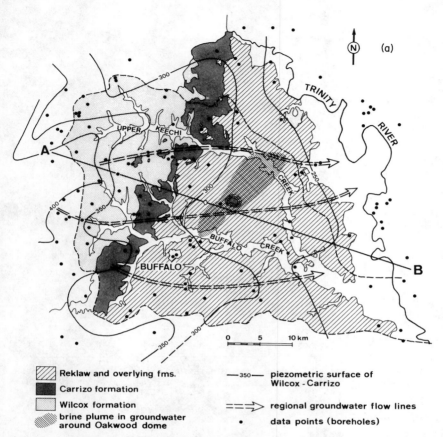

Fig. 65. Geological framework and hydrogeology of the Oakwood salt diapir, East Texas basin, United States. (a) Geological map showing the regional groundwater flow pattern in the Wilcox–Carrizo aquifer and the brine plumes indicating anomalous local flow conditions around the diapir. (b) Exaggerated vertical section of the diapir along line AB showing the surface recharge and discharge areas. (From Fogg and Kreitler, 1981, by permission of the Association of Engineering Geologists.)

internal structure which develops (Figs. 12 and 64) and the evidence from the pierced sedimentary pile indicates spasmodic movement, with rapid rise phases (rise velocities up to 15 cm/year) separated by long static periods. When the diapir reaches shallow levels, other processes begin to play an increasingly important role. Strong dissolution of the top and sides of the plug takes place in ground and formation waters, leaving insoluble residues of clay and gypsum as a caprock (see Figs. 12, 64, and 65). The caprock formations above many Gulf Coast salt diapirs indicate that many thousands of metres

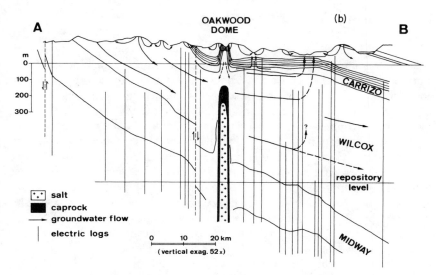

Fig. 65. Continued.

of salt have been dissolved to accumulate them. Many diapirs also show outward lateral spreading due to the reduction of the density difference and/or an increase in the rigidity of the near-surface rocks (Fig. 64), and movement may be stopped, to be restarted again later if renewed subsidence and sedimentation take place.

The complex geological histories of individual salt domes and diapirs are comparatively well known because of their association with oil and gas accumulations (intensive exploratory drilling) and because of extensive salt mining. In particular, the latter has provided valuable insights into the complexities of internal structure and into the very limited possibilities of structural prediction. Figure 64 shows the kind of structural detail which can be obtained from a limited number of exploratory boreholes in a previously unknown diapir (compare with the structure of an extensively mined dome, Fig. 12). The degree of structural prediction is one of the more critical points in assessing salt diapirs for use as repository hosts, since they often contain intricately folded and deformed layers of other evaporite minerals, which become unstable at rather low temperatures (e.g., carnallite, which dissociates at ~130°C) and must thus be avoided. A positive feature of many diapirs, however, is a much lower water content than bedded salts and, more generally, fewer impurities of any kind. This is related to the continual recrystallization of the salt minerals during diapirism, which tends to eliminate fluid inclusions from the individual crystals and to promote the removal of fluids from the whole salt mass.

IV. REGIONAL FLUID FLOW

The movement of groundwater at depth depends on the fluid pressure gradients and on the nature and distribution of aquifers (high-permeability zones), aquitards (low-permeability zones), and aquicludes (zero-permeability zones) in the bedrock. The first factor is determined by the regional topography, the second by the distribution and structural state of the different rock types. Clastic sedimentary sequences consist of different layers with different hydrogeological properties. The geometries of these are generally well known from extensive drilling, and the groundwater flow in any one formation can be modelled as a continuum using the well-established porous medium approach. The hydrogeology of crystalline complexes and other hard rock situations is generally less well known (less extensive drilling) and is generally dominated by flow along fractures. Numerical modelling under such conditions can make use of porous medium theory if the fractures are small and numerous relative to the rock volume under consideration and homogeneously or systematically distributed (see Chapter 10, Section II, for a discussion of the average hydraulic conductivity of the upper crust). However, more often than not crystalline complexes are heterogeneously jointed, with occasional large fracture zones directing the flow, and *in situ* testing yields wildly varying hydraulic conductivities between 10^{-1} and 10^{-11} cm/sec. Also, it has been shown that, under a given gradient, the amount of flow in a 100-m-thick sandstone horizon with a typical hydraulic conductivity of 10^{-5} cm/sec could be carried by a single fracture with an average aperture of 0.2 mm. Hence, there is a great hiatus between what is known and what is possible to know about the hydrology of sedimentary sequences and crystalline rock complexes.

At the present time, the hydrology of fractured media is perhaps the least undestood area of deep repository design and siting. Although the mathematics of calculating directional permeabilities from fracture orientation, spacing and aperture data has been developed using the parallel plate analogy for fracture flow, the basic data for its use are extremely difficult to obtain. The limits of relevant laboratory testing have been reached and *in situ* experimentation in boreholes and underground cavities is only just beginning. Field testing of fractured crystalline rocks has shown that some degree of prediction is possible on an empirical basis. However, many technical problems remain. Any kind of drilling, excavation and sampling tends to disturb the natural stress field, which in turn significantly alters fracture apertures and affects the permeability measurements. In addition, some types of field observation are still not technically possible, for example, the borehole measurement of fracture aperture and permeability anisotropy, and the determination in a borehole of whether or not a visible fracture is, in fact, a water pathway.

In contrast, the hydrology of sedimentary sequences is more manageable and investigation techniques have reached a high degree of sophistication in connection with subsurface oil, gas and water exploration. As an example, we take the regional flow pattern around a Gulf Coast salt dome which has been investigated as a potential site for deep burial of HLW.

Example: Oakwood Dome

The Oakwood salt dome is one of about 30 diapirs which penetrate Cretaceous and Tertiary formations in the East Texas Basin (Gulf Coast, the United States). The upper part of the diapir, together with its caprock, lies within the Wilcox-Carrizo Formation (Fig. 65), a 600-m freshwater aquifer which is tapped by about 1000 water supply wells throughout the basin. The Wilcox–Carrizo aquifer is semiconfined, underlain by > 1000 m of aquitards and aquicludes (separating it from the deep saline Woodbine aquifer, with its strongly depressed hydraulic head due to hydrocarbon production) and overlain by a thin aquitard, except above the dome and along the main outcrop (Fig. 65, pp. 200–201). It consists of alternating successions of sandstones and mudstones, with bed thicknesses normally in the range of 1–10 m and hydraulic conductivities in the range from 10^{-2} (clean sand) to 10^{-7} (mud) cm/sec.

Because of the exceptionally extensive well control, the regional flow pattern within the Wilcox–Carrizo aquifer is quite well known and proves to be topographically controlled. Water flow is down dip from the hilly region of the aquifer outcrop (recharge area) but leaks upward through the overlying aquitard towards the Trinity River, downstream from the dome. Fluid pressure versus depth relationships indicate that upward flow can extend as deep as 400 m beneath the Trinity River. Superimposed on this regional pattern is a dome-related perturbation which is less well known. The perturbation is revealed by a brackish water plume (see Fig. 65), which is so little salty that direct hydraulic communication between the salt and the groundwater can be ruled out. The plume extends northeast and southwest away from the dome, more prominently at lower levels in the aquifer, whereas the regional flow direction is towards the east. This curious effect is interpreted as due to a combination of local influences including a local topographic slope towards tributaries of the Trinity River to the northeast and southwest, local complexity of the aquifer framework due to dome-related faulting and distortion of beds, and exposure of the Wilcox–Carrizo aquifer in a small area immediately above the dome, creating a small local recharge system. The plume also suggests that radionuclides escaping from a repository deep within the dome would be significantly diluted within a distance of 10 km from the dome and would be unlikely to reach the surface. However, the processes of mixing, advective dispersion and retardation in such a heterogeneous aquifer are not yet well understood and detailed hydrological and geochemical investigations and numerical modelling are continuing.

V. PERTURBATIONS AROUND DEEP HLW REPOSITORIES

Up to this point, this chapter has discussed the physics of the upper few kilometres of the Earth's crust in its virgin state. Detailed knowledge of its physical state, and an assessment of future changes, natural or man-made, are necessary prerequisites for site selection and risk assessment. Future man-made changes arise from two main groups of activities: those connected with repository construction and operation, and those unrelated to radwaste disposal and possibly, in the distant future, carried out without knowledge of a repository's existence. Although the latter, human intrusion, is one of the major unknowns in long-term prediction (see Chapter 14, Section IV), here we briefly address the former question: what effects will a deep HLW repository have on the behaviour of the rocks and fluids in its surroundings and how significant are these effects likely to be? The main approach is through *in situ* testing and numerical modelling to full-scale prototype construction and long-time monitoring. At present, research and development is concentrated on field experimentation in several underground laboratories simulating actual repository conditions (Stripa mine, Sweden; Climax mine, Nevada Test Site, the United States; URL, Lac du Bonnet, Canada; Grimsel tunnel, Switzerland; Mol site, Belgium) and on complex three-dimensional modelling based on empirical data. Here, we review briefly some of the effects which have been observed or are expected and then take one example of a generic model, illustrating possible long-term developments (see also Chapter 14, Section III).

It is obvious that the construction of an underground cavity significantly alters the stress field in its surroundings, causing transient responses such as wall and pillar creep, movement on fractures, development of new cracks and fluid migration towards the cavity. Engineering geologists and mining engineers have been dealing with such problems in a whole range of geological situations for many years and have built up an enormous body of experience and expertise. This gives confidence that the short low-pressure phase of the life of a repository (construction, waste emplacement, backfilling, and closure) can be negotiated with a high degree of efficiency and safety. After this phase, the long-term evolution will be marked by a slow return to the virgin state of stress, confining pressure, and fluid flow field. At least, this would be the case if the emplaced waste were not heat-producing. In the case of HLW, however, the problem is compounded by the emanation of a thermal pulse from the waste cylinders, causing temperatures to rise to a maximum long after closure and significantly postponing a return to the virgin physical state. The thermal loading of a repository, usually given in terms of wattage averaged over the total area of the repository tunnel network, is thus an important parameter which in most radioactive waste management programmes is usually adjusted so that repository temperatures never exceed 100–150°C (by adjusting such factors as waste content in the canisters, above-ground storage time and canister spacing). Calculations for a salt repository at a depth of 1000 m show that such temperatures would be reached about 100 years after closure with a thermal load of 14 W/m² (as compared to a natural heat flux of ~ 0.05 W/m²). In dry rocks of lower thermal conductivity, maximum temperatures would build up more quickly and be more localized (e.g., at 500 m depth in granite, a loading of 10 W/m² results in maximum repository temperatures of 55–65°C after 20–50 years). A study which considered water-saturated welded tuff showed, as a further example, that with a loading of 25 W/m² and ambient rock temperature of 40°C, general repository temperatures (as opposed to canister internal temperatures) would be expected to reach ~ 120°C after 100 years and thereafter slowly decrease.

The thermal loading of the repository and its surroundings may cause several important changes in other conditions, thus triggering a whole series of interrelated physical (and chemical, see Chapter 11, Section III) processes. Thermomechanical effects may include crack formation due to differential thermal expansion of the host rock minerals, opening or closing of joints due to thermally induced compressive and tensile stresses around the repository, and a decrease in rock strength and increase in creep rate (in salt). Thermohydrological effects include the thermal disturbance of the regional groundwater flow system by convective energy transfer (see example below), changes in hydraulic conductivity due to thermomechanical effects (in fractured rocks), steam formation and/or buildup of abnormal fluid pressures (in shale), and the migration of fluid inclusions up the temperature gradient (in salt). *In situ* experimentation shows that these, and possibly other phenomena not yet identified, are interconnected in complicated cause–effect relationships which sometimes result in wide divergences of observed and predicted values in heater tests. For instance, thermal loading tends to reduce permeability around a repository, thus trapping fluids whose pressure increases as temperatures rise, but high fluid pressures reduce rock strength and facilitate brittle fracture, encouraging rock failure under the imposed thermomechanical stresses, in turn increasing permeability. Add to these complications other influences of radwaste emplacement, such as irradiation (e.g., hardening of salt under the influence of gamma radiation) or dehydration (e.g., release of bound water from clays, zeolites or evaporite minerals under temperatures above the dissociation point) and it is clear that until each individual process and its interrelationships is quantitatively understood, our predictive capacities are going to be rather limited. However, a consensus has been reached that maximum repository temperatures must be kept as low as reasonably achievable within the whole radwaste management system, preferably < 100°C.

Example: Thermohydrological Model

Figure 66 shows the results of a computer simulation of the evolution of a thermohydrological system, taking into account water properties (variations in density and viscosity with temperature), hydrogeologic properties of gran-

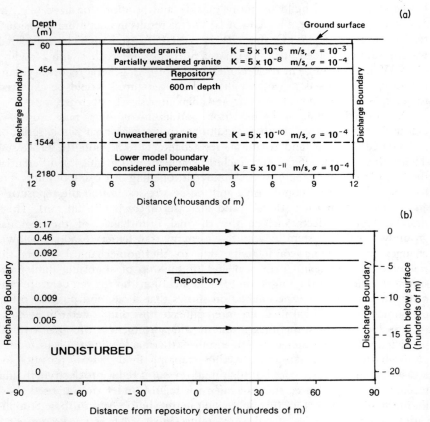

Fig. 66. Example of thermohydrological modelling to simulate possible convective effects around a HLW repository with a heat production of 25 W/m² at 500 m depth in granite (water table slope is 1:1000 from left to right): (a) the geological model with the hydraulic conductivities (K) and porosities (σ) of the different layers; (b) the calculated stream function contours of the undisturbed groundwater flow field; (c) stream function contours at approximately 10 years after repository decommissioning; (d) stream function contours at approximately 1000 years. (From Runchal and Maini, 1980, by permission of Pergamon Press Ltd.)

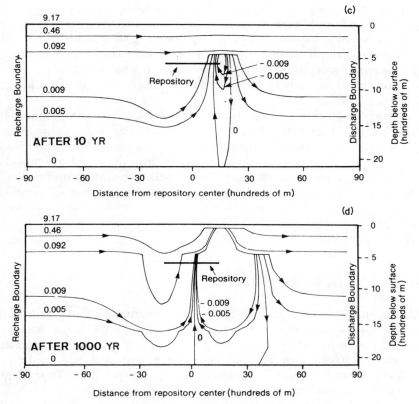

Fig. 66. Continued.

ite (typical values of hydraulic conductivity and effective porosity as a function of depth and degree of weathering), heat flow data (thermal conductivities, specific heats and normal terrestrial heat flow), idealized geological situation (depth of weathering, and regional groundwater flow pattern) and a conceptual repository design (depth, size and waste loading). The initial state shows layers of weathered granite (60 m), partially weathered granite (to depth 454 m), and unweathered granite with corresponding hydraulic conductivities and a water table slope of 1:1000, giving different flow rates at different levels (Fig. 66). A HLW repository is emplaced within this system at a depth of 600 m and with a thermal loading of ~ 25 W/m². This causes the slow development of a convective water circulation system downstream from the repository, which reaches its maximum effect (i.e., maximum upward water flux through the repository area) between 1000 and 4000 years after closure. During this period, upward flow rates of almost 4 m³/day are induced through the repository and water particles are estimated to take only

40 years to travel from the repository to the biosphere. Changing the geological boundary conditions does not appreciably alter the general order of magnitude of these numbers. This is an exploratory, generic model, not based on relations at a specific site. The major uncertainty lies in the hydrologic input data, in particular, in the hydraulic conductivities, which in a normal fractured granite are likely to vary strongly and unsystematically (see Chapter 10, Section IV).

It is important to note here that thermohydrological investigations and related numerical modelling have received considerable attention in recent years because of the surge of interest in geothermal energy. Numerous hot-water geothermal systems have been studied in great detail and will provide an important source of comparative data for HLW repository studies. Similarly, geological research into the genesis of hydrothermal mineral deposits (e.g., models of the thermal and fluid flow regimes around high level magmatic intrusions) may contribute significantly to our understanding of the long-term processes.

VI. CONCLUDING REMARKS

A deep-mined cavity containing HLW causes a perturbation of the physical conditions in the uppermost part of the Earth's crust. In this chapter we have explored the range and significance of these effects. A necessary prerequisite is a detailed knowledge of the subsurface geology, structural state, fluid flow field and stability of the site, i.e., a detailed site characterization. The problem of predicting subsurface relationships in sufficient detail is discussed further in Chapter 14. Assuming this can be overcome, there exists a broad base of practical experience, experimental data and sophisticated theory from the mining, construction and energy exploration industries to provide good assessments of the mechanical, hydrogeological and thermal effects of different types and sizes of underground repositories. Predictive capacity in this area is mainly limited by our inadequate understanding of the complicated feedback mechanisms operative in combined thermomechanical, thermohydrological and hydromechanical processes. The most promising way of tackling these is by field experimentation in repository-like underground test facilities.

SELECTED LITERATURE

Physical processes and rock properties in the upper crusts: Rutter, 1972; Fyfe *et al.*, 1978; Brace, 1980; Lappin, 1980; Schmidt *et al.*, 1980; Moore, 1981 (pp. 43–50); Ramberg, 1981; Robinson *et al.*, 1981; Grant *et al.*, 1982; Heard, 1982; Hsü, 1982 (pp. 95–127); Herrmann, 1983 (pp. 150–152); Heuze, 1983.

Brittle rock deformation—fracture system description and analysis: Grisak and Cherry, 1975; Sibson, 1977; Olkiewicz *et al.*, 1979; Thorpe, 1979; Carter *et al.*, 1981; Thorpe and Springer, 1981; Wilder and Patrick, 1981; Trask, 1982; Wittke, 1982; Brookins, 1983a (pp. 331–338); Swedish Nuclear Fuel Supply Co., 1983 (p. 18:1–163); Price, 1984.

Brittle rock deformation—faulting, fluid pressure and seismicity: Sibson *et al.*, 1975; Bell and Nur, 1978; Bolt, 1978; Paterson, 1978; Sägesser and Mayer-Rosa, 1978; Carpenter and Towse, 1979; Mayer-Rosa and Cadiot, 1979; Wyss, 1979; Kranz, 1980; Narasimhan *et al.*, 1980; Wahi *et al.*, 1980; Sibson, 1981; Rasmussen and Rohay, 1982; Swedish Nuclear Fuel Supply Co., 1983 (p. 8:1–49).

Ductile rock deformation—salt diapirism and internal structure of salt domes: Braunstein and O'Brien, 1968; Gera, 1972; Ledbetter *et al.*, 1975; Martinez *et al.*, 1975; Martinez and Thoms, 1977; Fairchild and Jenks, 1978; Mauthe, 1979; Bohn and Zoller, 1981 (pp. 139–158); Bornemann, 1982; Essaid and Klar, 1982; de Marsily and Merriam, 1982 (pp. 33–43); Carter and Hansen, 1983; Herrman, 1982 (pp. 128–203); Talbot and Jarvis, 1984.

Hydrogeology of porous media and regional sub-surface fluid flow: Getzen, 1977; Freeze and Cherry, 1979; Carr *et al.*, 1980; Narasimhan, 1980; Roberts and Cordell, 1980 (pp. 121–167); Fogg and Kreitler, 1981; International Atomic Energy Agency (IAEA), 1982b (pp. 635–668); Wang and Anderson, 1982; Brookins, 1983a (pp. 355–362); Greenkorn, 1983; Rosenheim and Bennett, 1984 (pp. 344–367).

Hydrogeology of fractured rocks: Kaufmann *et al.*, 1978; Marine, 1979; Brace, 1980; Runchal and Maini, 1980; Trimmer *et al.*, 1980; Witherspoon, 1980; Moore, 1981 (pp. 577–590); Davis, 1982; de Marsily and Merriam, 1982 (pp. 117–136); Topp, 1982 (pp. 199–230); Lutze, 1982 (pp. 509–558, 697–706); Swedish Nuclear Fuel Supply Co., 1983 (pp. 6:1–12, 14:1–21); Rosenheim and Bennett, 1984 (pp. 328–343); Kovari, 1983.

Temperature fields around deep HLW repositories: Rybach, 1975; Just, 1978; Llewellyn, 1978; Tin Chan *et al.*, 1978; Altenbach, 1979; Delisle, 1980; IAEA, 1980b (Vol. 2, pp. 105–222); Organization for Economic Cooporation and Development (OECD), 1981a (pp. 205–217); Wang *et al.*, 1981; Tarandi, 1983.

Thermomechanical and thermohydrological perturbations around deep HLW repositories: Office of Waste Isolation, 1976; Carter *et al.*, 1977; Norton and Knight, 1977; Hardy *et al.*, 1978; Wahi *et al.*, 1978; Norton and Taylor, 1979; Baca *et al.*, 1980; Claiborne *et al.*, 1980; Eaton *et al.*, 1980; Thunvik and Braester, 1980; Witherspoon, 1980; Runchal and Maini, 1980; Elder, 1981; OECD, 1981a (pp. 103–130, 205–229); Hofmann and Breslin, 1981 (pp. 87–170); Wang *et al.*, 1981; Eaton and Reda, 1982; Heuze *et al.*, 1982; de Marsily and Merriam, 1982 (pp. 101–115); Nelson and Rachiele, 1982; Topp, 1982 (pp. 355–362); Brookins, 1983a (pp. 515–538).

CHAPTER 11

Fluid–Rock Interaction

I. INTRODUCTION

Within the physical environment outlined in Chapter 10, a deep repository and its contents become part of a complicated chemical system. In a completely dry environment, reactions within the system would only take place by molecular diffusion in the solid state, and the geological evidence (e.g., the preservation of unaltered, high-pressure/high-temperature mineral assemblages at the Earth's surface) indicates that this is so slow that it is ineffectual over the time periods of interest here. However, equally compelling and widespread geological evidence underlines the importance of fluids as catalysts and agents of chemical reaction and transport in the Earth's crust, where they use fractured and/or porous zones as migration pathways. Hence, it is generally assumed that even an initially dry repository will eventually become affected by migrating fluids in the normal course of events. In this chapter, therefore, we look first at the geological evidence for fluid–rock interaction in the upper crust and try to assess the chemical environment into which the radwaste would be inserted when emplaced in a deep repository. In the following section, we consider a deep HLW repository after emplacement as a hot, interacting fluid–solid system of the same type, striving to attain thermodynamic equilibrium (near-field interactions). Finally, the leachates derived from this system are treated as radionuclide-bearing fluids which may enter the natural environment again, reacting with the rocks through which they flow and possibly transporting significant amounts of long-lived radionuclides to the biosphere (far-field interactions and geosphere transport).

II. CHEMICAL PROCESSES IN THE UPPER CRUST

The upper part of the Earth's crust is a chemical system consisting of a solid rock framework with heterogeneously distributed and partially inter-connected voids filled with aqueous solutions. Towards the Earth's surface, in the zone of weathering (see Chapter 5), the rock framework disintegrates, and the voids become larger and are partially filled with air. Passing down-wards into the lower crust, the zone of plutonic processes (see Chapter 9), the framework and voids coalesce into an incipiently ductile matrix or, locally, into rock melts. The thermomechanical and hydromechanical state of the upper crust (see Chapter 10) dictates that the mineral constituents and the chemistry of the enclosed fluids are continually adjusting to changing con-ditions of temperature, pressure, stress and fluid migration. Alteration, dis-solution and deposition of minerals by aqueous solutions become the dominant chemical processes and leave widespread evidence of their efficacy where they were allowed to continue over sufficiently long periods of time. In this section we look first at this geological evidence, particularly the hydro-thermal alteration of plutonic rocks and the formation of some types of ore deposits. This leads to a discussion of the present-day chemical makeup of deep crustal fluids, as deduced from direct sampling, and to problems of inter-pretation in terms of chemical evolution and fluid–rock interaction, especially in the case of deep groundwaters in crystalline complexes. The following sections then deal with the reaction of emplaced radwastes to this chemical environment.

A. Hydrothermal Alteration

Hydrothermal alteration refers to all the chemical changes which have taken place in rocks by the action of aqueous fluids under elevated temperature conditions (50–500°C). Some of these have already been touched upon., e.g., the diagenetic changes in sediments during deep burial, the alteration of vol-canic glass in contact with a mineralized vein, and the zeolitization of lavas and tuffs. Many of these changes are thought to improve some of the reten-tion properties of a potential host rock (e.g., to improve sorption capacity, reduce permeability or increase chemical stability) and can be expected to take place in the immediate vicinity of a deep HLW repository in the future.

Others may be deleterious (e.g., coating of fractures with nonreactive, less sorptive minerals or reduction of rock strength) and could adversely affect our capacity to predict subsurface relationships (e.g., by increasing heterogeneity).

Every rock shows secondary mineralogical changes which indicate a partial but pervasive alteration of the rock matrix by fluids circulating along cracks, cleavages, and grain boundaries, and through pore spaces. The secondary minerals are generally low- to intermediate-temperature forms (200–500°C) (as determined from fluid inclusions, element partitioning, and mineral stability fields). They are often hydrated and often show chemical or isotopic variations indicating the addition or subtraction of ions to and from the whole system. Microscopic study of plutonic rocks, i.e., those with primary high-pressure/high-temperature mineral assemblages, usually reveals a whole spectrum of such changes: olivine to serpentine; pyroxene to hornblende; hornblende, biotite and garnet to chlorite, limonite and/or ore minerals; and felspars to clay minerals, sericite and/or epidote. In sedimentary rocks, the difference between the primary mineral assemblages formed during diagenesis and the secondary effects formed under similar conditions at a much later date is much less obvious, even with quite different fluid chemistries, and the effects are less easily analysed. Hydrothermal alteration becomes more pervasive the more a rock undergoes deformation under hydrothermal conditions, and in contemporaneous shear and fault zones, new rocks are often formed which are made entirely of the corresponding hydrothermal mineral assemblages, with no remnants of the original mineralogy. In addition, dissolution tends to take place more readily in high-stress regions (pressure solution) and deposition in low-stress regions. Pressure solution effects include the truncation of mineral grains, fossil fragments or pebbles along contact surfaces marked by concentrations of otherwise dispersed impurities or along cleavage planes associated with intense folding (pitted pebbles, crenulation cleavage, stylolites, etc.) and (most spectacular of all) the formation of the caprock above salt diapirs (see Chapter 10, Section III,C). Low-stress situations in which mineral deposition and growth is dominant include the potential or actual voids created by tensional jointing and boudinage and the pressure shadows near rigid particles in a ductile deforming mass. From this wide field of geological phenomena, we take an example of a radwaste-related investigation to illustrate the principles involved.

Figure 67 shows the extent of hydrothermal alteration within one of the 1300 granitic plutons in the province of Ontario, Canada. The normally grey biotite- and hornblende-bearing granite shows gradations from grey to pink and grey to white, depending on the degree and type of alteration. The pink alteration is the most widespread and is seen in both surface outcrops and in the rock cores to be most intense in strongly fractured zones. The

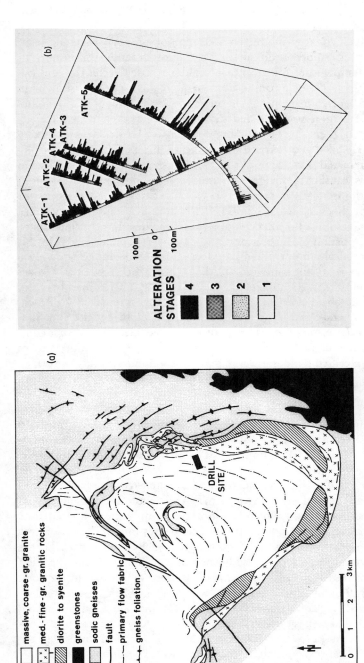

Fig. 67. The Eye–Dashwa Lakes granite pluton, near Atikokan, Ontario, Canada: (a) geological sketch map showing the typical petrography and structure of the pluton and the location of the exploratory drill site; (b) perspective view of the five drillholes showing the association of hydrothermal alteration with fracture intensity. Alteration stages: (1) no or very little alteration (grey granite), (2) low alteration, (3) medium alteration and (4) high alteration (deep pink granite). (From Kaineni and Dugal, in Bird and Fyfe, 1982, by permission of Elsevier Scientific Publishing Co.)

(a)

(b)

massive, coarse-gr. granite
med.-fine-gr. granitic rocks
diorite to syenite
greenstones
sodic gneisses
fault
primary flow fabric
gneiss foliation

DRILL SITE

N

0   1   2   3 km

ALTERATION STAGES
4
3
2
1

ATK-1   ATK-2   ATK-3   ATK-4   ATK-5

100m
0
100m

alteration involves changes in the composition of primary minerals (e.g., plagioclase becomes more sodium-rich), the growth and substitution of secondary low-temperature minerals (e.g., epidote, sericite, chlorite and calcite), a change in bulk rock composition (e.g., relative impoverishment in CaO and Sr and enrichment in MgO), and some development of disequilibrium among the various uranium and thorium isotopes present as trace elements, and among their short-lived daughters (e.g., ^{234}U deficiency with respect to ^{238}U). Detailed geochemical data indicate that fluids migrating along the fractures interacted with the wall rock by diffusion along microcracks and grain boundaries, causing exchange of chemical species, giving alteration products similar to the minerals in the fractures, and releasing some elements for transport out of the system in solution. The preliminary isotopic evidence indicates that migration of elements along these fractures has continued into the Pleistocene, possibly occurring over the past few thousand years. In this case, hydrothermal alteration has resulted in the altered granite showing a higher sorptive capacity for some radionuclides (e.g., ^{90}Sr), as well as a lower matrix permeability than the unaltered rock. Also, the mineral-lined fractures are expected to be more effective sorption surfaces than the naked wall rock, since the joint mineral assemblage is dominated by epidote, sericite and chlorite rather than calcite and prehnite. Experimental evidence suggests that complete hydrothermal alteration of silicate minerals in aqueous solutions takes place rapidly at elevated temperatures, e.g., in a few hundred years at $500°C$.

B. Hydrothermal Ore Genesis

Many ore deposits (economic concentrations of metals or metallic minerals) are classified as hydrothermal, i.e., as having formed from aqueous solutions at temperatures between 50 and $500°C$. The economic incentive to understand the genesis of such ores has led to intensive research, particularly over the last 25 years, and to the accumulation of an extensive data base on both ancient and modern hydrothermal systems. The whole area highlights an aspect of natural fluid action which has important implications for radwaste disposal, i.e., that its effect underground is not always one of dispersion and dilution of formerly concentrated constituents (usually assumed to be the main effect around a deep repository) but is sometimes one of concentration of formerly dispersed materials. For instance, in the ore-bearing quartz veins of the Yellowknife gold deposit (Fig. 68), where it has been shown that the ores were derived by lateral secretion from a 150-m-wide zone of sheared and now highly altered basic rock, concentration factors for the various elements were S:15, As:1125, Sb:1500, Cu:14, Zn:56, Au:1800 and Ag:4. Although the implications for fluid–rock interaction around deep HLW repositories have not yet been properly digested, we review below some of the main results of recent work on hydrothermal ore genesis, particularly with respect to the sources of hydrothermal fluids, the depth and time of formation, the sources of the dissolved constituents, and the problem of transport and precipitation.

Analysis of fluid inclusions inside transparent vein minerals such as quartz, sphalerite, or fluorite shows that in most cases the deposit-forming fluids were similar to those encountered in present-day geothermal systems and in deep aquifers. They were essentially brines with total dissolved solids varying between 30 and 500 g/litre, containing as major components Na, K, Ca and Cl. Mg, B, S, and Sr occurred constantly as minor constituents, and sometimes Fe, Zn, C and N were present in considerable amounts. The fluids were derived from one or several of the following sources: surface water (meteoric or oceanic), connate water (trapped surface water, diagenetic water, and/or deep groundwater), metamorphic water (from dehydration reactions), and magmatic water (from cooling of plutons or degassing of rising magma). Waters from the different sources should show distinctive $^1H/^2H$ and $^{16}O/^{18}O$ isotopic ratios, but actual measurements are often difficult to interpret because of mixing of waters from several sources or long-continued fluid–rock interaction. The forces which drive a fluid through its channelways are also diverse, and difficult to reconstruct unambiguously in the case of now-inactive systems. However, indirect evidence indicates that the duration of mineralization in many cases was < 1 m.y. and in some cases not more than a few thousand years. Similarly, circumstantial data lead to the conclusion that most hydrothermal deposits formed in the upper 3 km of the crust.

The sources of the dissolved constituents (later deposited as ore minerals) is rarely definable and a hydrothermal fluid must be regarded as an evolving entity, possibly containing late-magmatic species and certainly containing material dissolved from the rocks through which it has passed, dissolving and depositing minerals according to changing conditions of temperature, pressure, rock environment, and degree of mixing. In the case of lead minerals, lead isotope ratios can be used as distinctive fingerprints for their origin. This has revealed a surprising diversity of origins. Different examples indicate release in magmatic fluids, leaching of basement complexes, and transfer to the solution by diffusion out of solid solution in detrital felspars, and the geochemical cycle of lead is becoming increasingly well understood. It is assumed that hydrothermal fluids collect all metals in the same ways, and the lead isotope data suggest that the main process is channelway reaction. Any rock can serve as a source of geochemically scarce metals under certain conditions, in some cases (Pb and Cu) without any special enrichment, in others, apparently, only if it itself is already enriched (Sn, Ag and Hg). That hydrothermal ore deposits are generally scarce and small, however, is mainly due to the insolubility in water of the more common ore minerals, particularly

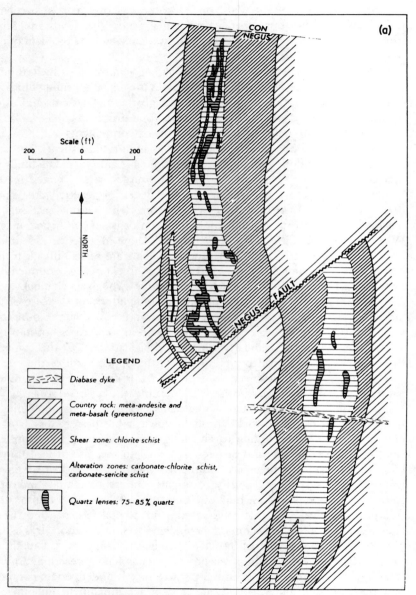

Fig. 68. Example of a hydrothermal ore deposit: the gold–quartz veins in the Yellowknife District, Northwest Territories, Canada. (a) Geological map of the 1775-ft level in Negus Mine showing one of the 100- to 150-m-wide shear zones in which the alteration and mineralization has taken place; (b) composite diagram of the chemical changes across the shear zones (averaged over the whole district). (From Wanless *et al.*, 1960, and Boyle, 1959, by permission of the Society of Economic Geologists.)

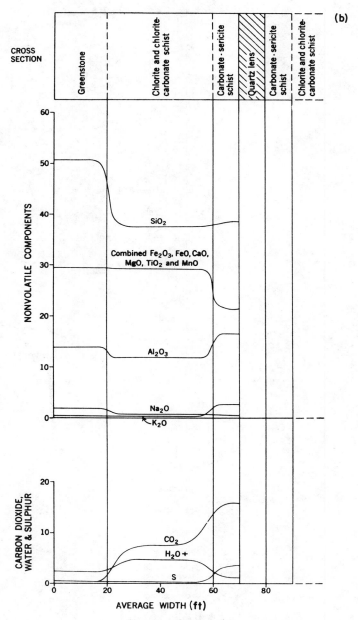

Fig. 68. Continued.

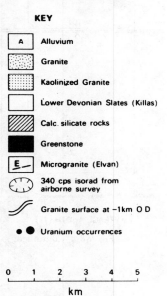

195   200   205

065

060

055

St AUSTELL BAY

N

KEY

A	Alluvium
	Granite
	Kaolinized Granite
	Lower Devonian Slates (Killas)
	Calc. silicate rocks
	Greenstone
E	Microgranite (Elvan)
	340 cps isorad from airborne survey
	Granite surface at −1km O D
● ●	Uranium occurrences

0   1   2   3   4   5

km

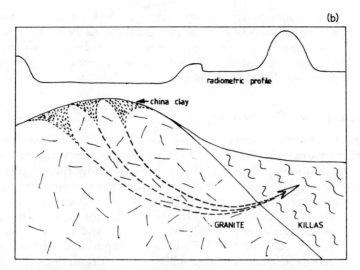

Fig. 69. Example of hydrothermal redistribution and concentration of uranium. (a) Geology of the St. Austell granite, Cornwall, United Kingdom, showing associated kaolinized zones, radiometric anomalies, and uranium occurrences (pitchblende veins). (b) Diagrammatic section illustrating the proposed mechanism of uranium mineralization during Tertiary time: magmatic uraninite in the granite is leached by oxidising fluids related to kaolinization and supplies uranium to a convection cell, leading to pitchblende deposition in restricted fractures of low Eh in the country rock (Killas). (From Ball *et al.*, in International Atomic Energy Agency, 1982c, by permission of the International Atomic Energy Agency.)

sulphides. The transport of most metals in aqueous solution depends on the formation of complex ions, of which two are particularly important: chloride complexes (e.g., $ZnCl_2$ and $CuCl_3^-$) in chloride-rich solutions and reduced sulphur species [e.g., $HgS(H_2S)_2$ and $Zn(HS)_3^-$] in sulphur-rich solutions. Other ligands such as fluoride, hydroxide and sulphate, and even organic complexes (see Chapter 5, Section III,F) may play a significant role locally. Precipitation from such solutions to give the actual ore concentration depends on localized conditions which cause deposition at the same location over a long period of time (a change in temperature, pressure, wall rock composition, and/or fluid composition due to mixing).

This short subsection does no justice to the vast amount of work which has been carried out on hydrothermal ore deposits. Particularly important for the problem at hand is an understanding of vein-type uranium deposits (Fig. 69), which are found in a wide range of rock types, generally in specific provinces (such as magmatic, pegmatitic, and sedimentary) in which other uranium mineralizations also occur. Age determinations on several deposits within a specific province indicate repeated concentration, transportation and redeposition over long periods of time, involving both surface and subsurface processes as integral parts of the uranium geochemical cycle. The increasing exploration for new uranium resources to provide nuclear fuel will no doubt greatly increase our knowledge of natural processes which are also relevant to the back end of the nuclear fuel cycle.

C. Groundwater Geochemistry

In the last subsections, we have considered hydrothermal alteration and hydrothermal ore genesis as natural analogues of processes which are expected to take place around a deep HLW repository. We now turn to present-day fluids in the crust in order to define the chemical environment into which such a repository would be placed. The chemistry of deep groundwaters, as determined by the analysis of waters from boreholes, mines, wells, and springs, is dominated by the cations Ca^{2+}, Mg^{2+}, Na^+ and K^+ and the anions HCO_3^-, SO_4^{2-}, Cl^- and NO_3^-. Cation compositions are, as indicated above, extremely variable and dependent on local conditions. In contrast, anion compositions in many sedimentary basins show a general evolutionary trend which gives a rough measure of water maturity (i.e., age, length of flow path and/or depth). Shallow groundwaters in zones of active flushing through well-leached rocks tend to be low in total dissolved solids (TDS) and to be relatively rich in HCO_3^- (from passage through the CO_2-rich soil zone and from dissolution of limestone and dolomite). At intermediate depths or distances, in zones of less active circulation, the TDS concentration is higher and SO_4^{2-} becomes the dominant anion due to dissolution of sulphate minerals along the flow path and to redisposition of calcite (removal of HCO_3^- also causing a change from acid to neutral or slightly alkaline). In deep zones of very sluggish groundwater flow, high TDS and high Cl^- concentrations become characteristic and are thought to be due to the dissolution of occasionally encountered halite beds and the incorporation of traces of trapped seawater along the very long flow path.

Parallel to this compositional trend (reflected in the trend from low to high pH), two other trends have been distinguished in sedimentary rock sequences: in redox potential and in isotopic ratios. The evolution from high to low redox potential (from oxidising to reducing conditions) is essentially determined by the rate of oxygen introduction in rainwater from the surface and the rate of oxygen consumption by the bacterially mediated decomposition of organic matter and the action of redox buffers such as Fe $(OH)_3$ and Fe_2O_3 (see Chapter 5, Section III,D). These factors vary greatly from aquifer to aquifer, but the slower the groundwater circulation (the longer the residence time), the lower the resulting redox potential, so that groundwaters at great depth are usually free of dissolved oxygen. Most sedimentary basins also show a steady enrichment of ^{18}O in the groundwaters compared to the $^{18}O/^{16}O$ ratio of meteoric water in the corresponding recharge area. As the water flows deeper, fluid–rock interactions intensify under higher temperatures and pressures, and since sedimentary rocks are typically enriched in ^{18}O relative to meteoric water, the groundwater ratio is raised. Combining $^{18}O/^{16}O$ ratios with hydrogen isotope ratios for samples from different levels in a deep sedimentary groundwater system generally yields a very pronounced pattern which is characteristic for the system and gives clues as to its particular long-term evolution.

These trends have been recognized in all major sedimentary basins (data from oil and gas exploration) and in all hot water–steam fields (data from geothermal energy exploration), but there is as yet little information from crystalline complexes. What there is suggests a different type of evolutionary pattern. As an example, we take a recent study of the problematic saline deep groundwaters of the Canadian shield (Fig. 70). These occur in many different types of crystalline rock and are thought to be characteristic of depths > 600 m. The salinity often exceeds 200 g/litre and the chemistry of the brines (e.g., high Ca^{2+}, low SO_4^- and different Br/Cl and Br/I ratios) is such as to distinguish them strongly from evaporated seawater (e.g., low Ca^{2+}, high Mg^{2+} and SO_4^-) and from shallow groundwaters from the same localities (e.g., low salinity, low Ca^{2+}, and HCO_3^- content increasing with time, as expected for the evolution of meteoric water by low-temperature reaction with minerals in crystalline rocks). Stable isotope analysis (Fig. 70) shows that the deep saline groundwaters lie on lines above the global meteoric water line, in contrast to the lines for shallow groundwater from the same locality and for deep sedimentary basins, hydrothermal fluids and ore-forming solutions the world over. The processes which led to these anomalous chemical and isotopic compositions are not yet understood, but they are thought to include an initial penetration of the shield basement by saline waters, the addition of the juvenile component from deep-seated metamorphic processes, and subsequent modification by fluid–rock interaction in a closed system over a long period of time. At least, the evidence indicates that they were not generated in modern environments and are probably many millions of years old.

A detailed description of the chemistry of the groundwater and its variation above and below repository level is an indispensable element in any site characterization. The acidity (pH) and the redox conditions (Fig. 71) strongly influence such processes as canister corrosion, leaching of vitreous or ceramic waste forms, dissolution of uranium oxide fuel, speciation of the various radionuclides in solution, sorption and transport. Particularly critical from the point of view of modelling such processes is, in this respect, the difficulties in the field measurement of redox potentials. Other parameters which affect

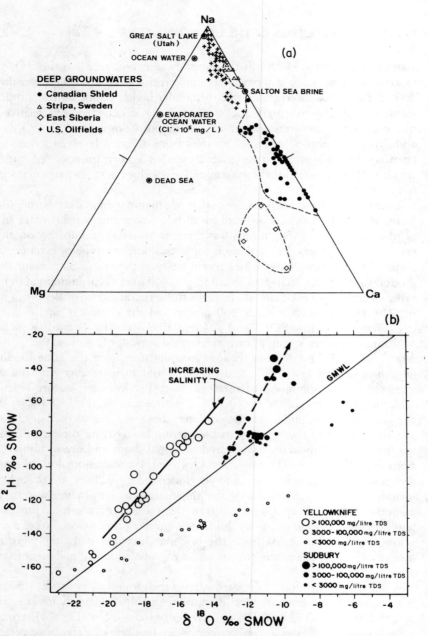

Fig. 70. Saline deep groundwater from the Canadian shield. (a) Na–Mg–Ca (mole%) ratios of brine and saltwater (> 10⁴ mg/litre TDS), compared with brine and saltwater from other environments. (b) δ^2H and $\delta^{18}O$ values of surface groundwater (generally < 3,000 mg/ litre TDS) and deep saline groundwater (generally > 100,000 mg/litre TDS) from the Yellowknife and Sudbury mining areas, showing the different isotopic compositions (different sides of the global meteoric water line, GMWL; standard mean ocean water, SMOW). (From Fritz and Frape, in Bird and Fyfe, 1982, by permission of Elsevier Scientific Publishing Co.)

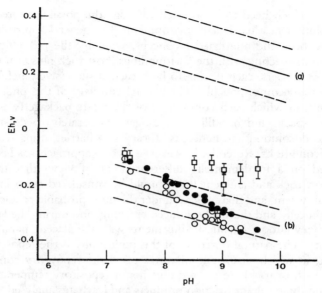

Fig. 71. Measured and calculated Eh/pH values for deep groundwaters from the Baltic Shield, Sweden. Line (a) gives the relation for open systems in contact with atmospheric oxygen, line (b) for closed systems in the presence of the redox components Fe(II) and Fe(III). Calculated Eh/pH data from the deep groundwaters are based on measured Fe(II) concentrations and pH, under the assumption of equilibrium with goethite, FeOOH, (○) or goethite and siderite, FeCo₃ (●). That the direct Eh measurements (□) do not agree with the calculated data reflects the experimental difficulties in the field measurement of redox potentials. (From Swedish Nuclear Fuel Supply Co., 1983.)

these processes to varying degrees are the content of corrosive substances, the presence of complexing agents (such as hydroxides, carbonates or organic ligands), and, for certain radionuclide systems, the total salinity. With regard to the latter, characterization of the deep saline groundwaters described above is of equal importance since there is a high probability that a repository at a few hundred metres depth in a shield area will spend some or most of its life in such a geochemical environment (see Chapter 13, Section IV,B).

III. RADWASTE DISSOLUTION: NEAR-FIELD PROCESSES

It is generally agreed that the most likely scenario of radionuclide release from a deep repository is by dissolution in the deep groundwaters followed by migration of the radionuclide-bearing solutions through the geological environment (geosphere) to the biosphere. Interactions between fluid, rock and waste in and immediately around the repository, called near-field processes, are the ones which need to be assessed to judge the possible degree of rad-waste dissolution in a particular environment. The general approach to this problem has been the multibarrier concept, whereby the radionuclides are prevented from reacting with the surroundings (host rock plus groundwater) by a series of barriers, each having to be breached successively before radio-nuclide release becomes possible. The natural corollary of this philosophy is the leach test, in which each component of the waste package (waste form, canister, over-pack, and backfill) is investigated separately to determine its resistance to dissolution and hence its efficacy as a barrier, using realistic ex-perimental conditions. Recently, however, a second approach has been intro-duced based on a fundamentally different point of view: that the waste package, host rock and ambient fluid should be considered as an interactive geochemical system, and that, at least under the most probable physical failure mode (tectonically and thermally induced cracking) and during the late stages of repository evolution, chemical adjustments will take place in parallel rather than in series. The natural outcome of this philosophy is the repository sim-ulation test, in which small-scale mixtures of all the repository components (including the host rock) are reacted together at repository temperatures and pressures and the resultant reaction products and leachates analysed. The two approaches are at the same time complementary and contradictory. Although both contribute to the production of waste packages with materials and di-mensions which ensure extremely slow dissolution rates, a capability of quan-titative prediction of maximum release rates under deep repository conditions (i.e., the "source term" for far-field models) has still to be developed (see Section IV).

A. Leaching

The main processes which determine the rate of release of radionuclides and other elements to the surrounding solutions are diffusion-controlled ion exchange, surface corrosion, and chemical reconstitution. Ion exchange takes place as cations diffuse to the surface and are taken up into solution (being replaced by hydrogen or hydronium ions) or into colloids, or are deposited as hydroxides. The overall result is the development of an amorphous, hy-drous, silica-rich surface film, which slows down the rate of release as it be-comes wider (Chapter 8, Section IV,A). This is a process of incongruent dissolution, since each element is released at a different rate. Surface corrosion of the silica-rich surface film takes place by chemical dissolution, mechanical flaking or microbial activity and is a process of congruent dissolution, usually becoming dominant only after the buildup of the surface film. Under chem-ical reconstitution are placed all solid-state reactions within the glass and sur-face film which lead to the development of new phases in equilibrium under the prevailing conditions. Such phases often retard outward diffusion by sorption processes. The first two processes dominate at low temperatures because of the sluggishness of solid-state reactions. As temperatures increase, all reactions tend to take place more quickly, the first two generally increasing the leach rate and the last generally decreasing it. The higher the temperature, the more rapidly diffusive processes take place, until a situation is reached in which solutions, residual solids, and hydrothermal alteration products are continuously reacting in order to maintain thermodynamic equilibrium. It is clear that leach tests involving just one solid and one liquid phase cannot give a definitive picture of these combinations of processes, although they do allow the relative stabilities of various waste forms and package materials to be determined under a variety of conditions.

B. Complex Reactions

The near field around a young HLW repository is best considered as a complex geochemical system of the type which has been investigated by experimental petrologists over many decades. Repository temperatures during the time period in which the waste contributes significant amounts of heat vary between 100 and 300°C, depending on design concept and host rock, and fluid pressures of 100–300 bar are expected, with estimates in some scenarios up to 800 bar. In addition, radiolysis, the splitting of water into hydrogen and oxygen and/or oxidizing compounds (e.g., H_2O_2) under the influence of ionizing radiation, is expected to be important at this stage (Fig. 72). It is of particular concern because it produces an oxidising environment and an outward-migrating redox front around each HLW can-

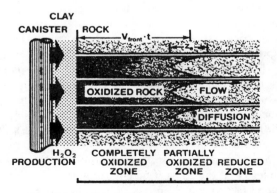

Fig. 72. Propagation of the radiolysis front in the direction of groundwater flow away from a HLW canister according to the Swedish concept. (From Swedish Nuclear Fuel Supply Co., 1983.)

ister, within which several important radionuclides (e.g., U, Np and Tc) are much more soluble. Although experiments with mixtures of waste form–waste package materials, host rock, and groundwater to simulate such conditions are as yet of a preliminary nature, it has been shown that radionuclides that were extracted into solution in leach tests are often immobilized in new minerals. Borosilicate glass may become extensively altered under such conditions, but the alteration products (e.g., acmite, pollucite, orthoclase, analcime, or weeksite) act as scavengers, fixing Cs, Sr, and other radionuclides in mineral form. Conversely, some experiments with ceramic waste form minerals such as perovskite or hollandite (see Chapter 9, Section IV) show that they are unstable and highly leachable in groundwaters with typical compositions in the presence of basalt or shale. Repository simulation experimentation, although only in its infancy, promises to provide an important foundation for the quantitative prediction of dissolution and reaction processes in the very complex near-field geochemical system.

IV. RADIONUCLIDE MIGRATION: FAR-FIELD PROCESSES

Fluids produced by the chemical processes in and immediately around the repository are expected to migrate outwards under the influence of the perturbed, regional groundwater flow system. There, they may enter a chemical environment quite different to that in the repository: rocks of different and changing composition, fluids of different and changing chemistry and surroundings of different and changing conditions (temperature, pressure, etc.). As in the case of migrating leachates in the weathering zone (see Chapter 5), the retarding properties of the rock framework are the main factors preventing radionuclide migration in this far-field zone. Recognition of this led to extensive laboratory testing of various materials (rocks, backfill materials, and alteration products) in the form of now standardized sorption tests. In these tests, the distribution coefficient is determined for a particular ion between its solution in a fluid of the required chemistry and powdered or pelletized aggregates of the solid in question. These data and other information on sorptive behaviour are now stored in a central data bank at Saclay, France [International Sorption Information Retrieval System (ISIRS)]. However, just as with leach tests, the value of sorption test data for predictive modelling is questioned by many earth scientists because the test conditions in no way reflect the complexity of the subsurface environment and because sorption is only one of several factors which influence geosphere transport. This has led to a different approach, empirical field experimentation, in which the aim is to simulate actual subsurface conditions (e.g., in underground laboratories) or to study the effects of other radionuclide releases into the bedrock (e.g., around underground nuclear bomb tests). A corollary of this approach is the intensive search for migration data in natural geological situations and the study of the contamination around natural repository analogues such as uranium ore deposits or the natural reactor phenomenon at Oklo, Gabon (see below).

A. Retardation Mechanisms

From natural and man-made field experiments, from extensive laboratory testing and from theoretical studies, it has long been known that all important radionuclides move more slowly through a rock matrix than do the fluids carrying them in solution. The ratio between the velocity of the water and the velocity of the radionuclide in a given environment is called the retardation or retention factor, R. Retardation can be thought of as due to two main processes, sorption and matrix diffusion, of which the former is usually the most important. Sorption is a general term covering several physico-chemical phenomena, such as adsorption, ion exchange and isomorphic substitution, which define types of fixation of ions on fluid–solid interfaces and whose effects are difficult to keep strictly separated. Matrix diffusion designates the penetration of ions into dead-end micropores or microcracks in the rock matrix, thus effectively removing them from the main fluid flow field. This process even retards radionuclides which are nonsorbing, like iodine or tritium. For most radionuclides, however, its effect is to increase the retardation effect due to sorption, as expected from the results of sorption tests.

Laboratory sorption tests are of various types, but the most common is the batch, or static equilibration, test in which the solute containing the radionuclide is allowed to reach equilibrium with the solid of interest. For equilibrium conditions, there should be a straight line relationship between the concentration of sorbed radionuclide and the concentration of the same ion in solution expressed with respect to either the solid weight–fluid volume ratio (K_d) or the solid surface area–fluid volume ratio (K_a). K_d is easily measured but is obviously a less easily interpreted parameter than K_a, which is, however, very difficult to determine accurately because of its dependence on surface area. The equilibrium condition is thought to be relevant to the natural situation because very slow groundwater flow is one of the main repository siting criteria, and the temperatures used in most tests (20–25°C) are similar to those expected in the upper 1 km of the crust.

Some divergence from a straight line relationship has been observed, indicating a concentration-dependent partitioning which probably reflects the different importance of the various sorption phenomena under different conditions. Physical adsorption is a loose fixation of ions on the surface of any solid particles due to surface tension forces. Ion exchange designates a process of fixation and release which depends on the intrinsic properties of the minerals involved. Clay minerals and zeolites are the most important ion exchangers in subsurface environments, but many other minerals which occur in fine-grained aggregates may be very efficient. Being also a surface phenomenon, ion exchange is reversible; adsorption and desorption take place rapidly in response to changing conditions. In contrast, isomorphic substitution is almost irreversible at low temperatures. It is a diffusion mechanism by which ions in the solution are substituted for ions of the same charge and ionic radius in the crystal lattice of the solid. The relation between these phenomena is not known in detail and much experimental work will be needed to clarify this issue.

In addition to these retardation mechanisms, several other factors affect radionuclide migration through the geosphere. The complexities of the far-field chemical system and the changing chemical parameters along the flow paths do not need further description. Here, we note simply two important effects which work against retardation, i.e., transport-enhancing mechanisms. The first is the prevention of sorption on the surfaces of the rock matrix by the formation of soluble complex molecules (in deep groundwater mainly carbonate and organic complexes) or by sorption on colloidal particles in the fluid (see Chapter 5, Section III,F). The second is dispersion, which is the collective term for the mechanisms by which a dissolved substance spreads out to larger and larger liquid volumes as the liquid flows in a medium. Although this process causes dilution, it also results in a small proportion of dissolved radionuclides being transported more rapidly than the average (channelling). The unexpectedly rapid transport of small amounts of sorbing tracers in recent field experiments and the sometimes poor agreement between predictions based on laboratory data and actual measurements may be partly due to a combination of the above effects.

B. Natural Migration Analogues: Oklo

Of all the natural evidence for the extent of migration of dissolved or diffusing substances in rock matrices, the most relevant to radwaste disposal is the natural reactor system found in the uranium mine at Oklo, Gabon (Fig. 73). The uranium mineralization in the Oklo mine occurs at a particular horizon (C-1) in the lowermost formation (F_A) of the Francevillian Series, a succession of unmetamorphosed, middle Precambrian volcanic and sedimentary rocks. Horizon C-1 is 5–6 m thick, a sequence of conglomerates and impure sandstones ($\sim 10\%$ clay minerals) with occasional argillaceous interbeds. Isotopically normal uranium ore (pitchblende) occurs either as disseminations in highly recrystallized black sandstones (up to 1% UO_2) or concentrated in irregular zones of fractured rock (5–25% UO_2). Within this

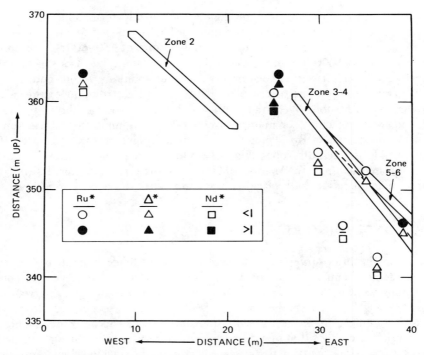

Fig. 73. Redistribution of ruthenium isotopes (Ru*) technetium-99 ($\triangle$*), and neodynium isotopes (Nd*) around reactor zones 2–6 at the Oklo mine, Gabon. Open symbols represent relative deficiencies of elements, closed symbols relative excesses. (From Curtis *et al.*, in Hofmann and Breslin, 1981).

layer, 13 reactor zones, which show several anomalous features, have been identified. Within the zones, sedimentary features disappear and there is an almost complete lack of quartz. Total silica content decreases from ~ 90% in the normal sandstone to ~ 35% in the reactors, quartz being replaced by clay minerals showing progressive mineralogical changes in the aureoles around the zones. The uranium content increases to 25–60% UO_2 in the form of coarsely crystalline uraninite, which is isotopically anomalous (e.g., ^{235}U content significantly lower than the normal value of 0.72%) and which seems to have remained thermally and mechanically stable since the reactors were functional, an estimated 2000 m.y. ago.

At that time, the uranium ores became locally so concentrated that they went critical, a condition made possible by the then higher ^{235}U content (~ 3%, approximately the enrichment required to produce fuel for modern light water reactors). Detailed isotopic investigations indicate that the duration of criticality was around 300,000 years and that about 12 tons of ^{235}U

were "burnt up" in that time, producing a thermal load of about 50 W/m² (two to five times greater than the thermal load envisaged in HLW repositories). Fluid inclusion studies indicate temperatures of 450–600°C and pressures of up to 1000 bar during the fission period, and other data suggest the picture of convectively circulating hot aqueous solutions within and up to 30 m outside the zone of criticality. The spatial redistribution of the now remaining, very long-lived transuranics and fission products shows the following pattern: actinides remained fixed within the uraninite crystal structure; fission products like Ru, Tc, and Nd or their daughters migrated up to 10 m outside the reactor zones (Fig. 73); and fission products like Cs and I disappeared completely. Analysis of Ru–Tc isotopic relationships has shown that their migration took place within 1 m.y. after the end of criticality.

V. CONCLUDING REMARKS

From this brief survey of the chemical action of fluids in the Earth's crust and of the chemical impact of a deep HLW repository, some general conclusions seem to be warranted. First, detailed groundwater characterization studies will be necessary before an asessment of the geochemical effects of radwaste emplacement at a specific site can be made. Second, the chemical and isotopic composition of a deep groundwater can help to elucidate its past evolution, which in turn permits prediction of the average, long-term hydraulic properties of the site. In the case of crystalline rocks (fractured media), this information is difficult to obtain in any other way. Third, natural hydrothermal systems can change character considerably on a time scale which is comparable to the nominal isolation time of HLW, sometimes much less. However, such changes are likely to remain within geochemical limits which are slowly becoming more clearly defined. Fourth, natural analogues for near-field and far-field processes abound in the Earth's crust. The only natural nuclear waste repository known, Oklo, demonstrates that the migration of many radionuclides in some situations is negligible over geological time periods. In contrast, other geological situations, e.g., those leading to the formation of vein-type uranium ore bodies, apparently favour migration over large distances and the action of problematic concentration processes.

SELECTED LITERATURE

Hydrothermal alteration and ore genesis: Boyle, 1959; Wanless *et al.*, 1960; International Atomic Energy Agency (IAEA), 1974 (pp. 515–604), 1982c; Park and MacDiarmid, 1975; Fyfe *et al.*, 1978; Barnes, 1979; Evans, 1980; Organization for Economic Cooperation and Development (OECD), 1981a (pp. 93–101); Crisman and Jacobs, 1982; Bird and Fyfe,

1982 (pp. 35–58, 87–102); LeAnderson *et al.*, 1983; Maynard, 1983 (pp. 147–180); Wood and Walther, 1983.

Groundwater geochemistry in the bedrock: Freeze and Cherry, 1979 (pp. 237–302); Benes and Majer, 1980; Hobson, 1980 (pp. 139–188); Neretnieks, 1981; Rydberg, 1981; Bird and Fyfe, 1982 (pp. 179–190); Perry and Montgomery, 1982; Sanford, 1982; Allard *et al.*, 1983; IAEA, 1983e,f (pp. 273–422); OECD, 1983c (pp. 133–172); Swedish Nuclear Fuel Supply Co., 1983 (p. 7:1–21); Wikberg *et al.*, 1983.

Geochemistry of near-field processes (radwaste dissolution): McCarthy *et al.*, 1978; Westsik and Turcotte, 1978; Fried, 1979; OECD, 1980b (pp. 67–74); 1981a, 1982c; Northrup, 1980 (pp. 445–600); Moore, 1981 (pp. 131–138); Johnston and Palmer, 1982; Bird and Fyfe, 1982 (pp. 59–86); Roy, 1982 (pp. 124–146); Topp, 1982 (pp. 295–456); West *et al.*, 1982; Swedish Nuclear Fuel Supply Co., 1983 (pp. 12:1–29, 13:1–29).

Geochemistry of far-field processes (geosphere transport): De Marsily *et al.*, 1977; OECD, 1979b; 1982e; Dosch, 1980; Neretnieks, 1980; Northrup, 1980 (pp. 601–664); Hofmann and Breslin, 1981 (pp. 171–254); Moore, 1981 (pp. 457–514); Neretnieks, 1981; Rasmuson and Neretnieks, 1981; Bird and Fyfe, 1982 (pp. 123–138, 191–214); Jensen, 1982; Luze, 1982 (pp. 679–848); Topp, 1982 (pp. 187–294); Walter, 1982; West *et al.*, 1982; Swedish Nuclear Fuel Supply Co., 1983 (pp. 12:1–29, 14:1–32); Landström *et al.*, 1983; Moreno *et al.*, 1983.

Oklo and other natural migration analogues: IAEA, 1975; 1982b (pp. 603–612); Brookins, 1976; 1983b; Cowan, 1977; McCarthy, 1979 (pp. 355–366); Gore *et al.*, 1981; Hofmann and Breslin, 1981 (pp. 255–283); Crisman and Jacobs, 1982; Granath *et al.*, 1983; National Academy of Sciences (NAS), 1983 (pp. 300–305); Curtis and Gancarz, 1983.

CHAPTER 12

Ocean Processes

I. INTRODUCTION

In this chapter and part of the next, we turn to processes at work in the Earth's oceans. This chapter is concerned with the use of the ocean floor and the ocean sediments for radwaste disposal, Chapter 13 with their use as an indicator of past and future climatic change. In both, we tie together loose threads left in the course of earlier discussions: ocean disposal concepts in Chapter 2, the water cycle in Chapter 3, and stable sedimentary environments in Chapter 7. Here, we first review some physical, chemical, and biological aspects of ocean processes, particularly with respect to the deep water reservoir. This leads to a consideration of deep ocean sediments and their vertical and lateral variations as exemplified by an area in the northwest Pacific which has become well known from extensive deep-sea drilling and coring. In a final section, we consider radioactivity in the ocean in general, and particularly its possible release from HLW canisters emplaced in the seafloor sediments as envisaged in the subseabed disposal concept.

II. OCEAN PHYSICS AND CHEMISTRY

The mass of ocean water shows a physical and chemical differentiation into a surface water layer (down to 200–300 m, comprising $\sim 2\%$ of the ocean's volume) and a deep-water mass (below 500–1000 m, comprising

~ 75% by volume), separated by a transition zone known as the thermocline. As the name implies, the thermocline marks a zone of rapid change in temperature from values which reflect climatic conditions in the surface layer to values of $< 5°C$ in the deep water mass. Hence, the thermocline is sharpest in the tropics and may disappear entirely in polar regions, at least during the cold season. The thermocline is also a zone of relatively rapid increase in density (in the North Pacific and North Atlantic, from < 1.025 above to > 1.027 below) and decrease in salinity (to a constant value of 34.4–36.0‰ in the deep water mass). The surface layer roughly corresponds to the water layer penetrated by light (100–200 m in the clearest subtropical seas) and the thermocline roughly corresponds to the sound channel (the region between 400 and 900 m in which the velocity of sound waves decreases with depth). In general, the thermocline constitutes both a physical and chemical barrier, and the ocean can be considered as a two-reservoir system in many areas, with only limited interconnection. Knowledge of the dynamics of the surface layer is important in assessing the effects of releasing radioactive materials directly into the surface waters as is done at the present time from the fuel reprocessing plants at Windscale (United Kingdom), La Hague (France) and Trombay (India) and to a much lesser extent from nuclear-powered ships. The dynamics of the deep-water mass are important for evaluating the long-term effects of LLW dumping as practised until recently by the OECD Nuclear Energy Agency (see Chapter 2, Section III,A) and of subseabed disposal as proposed by the international Seabed Working Group (see Chapter 2, Section IV,F and Chapter 15, Section III).

Conditions in the surface layer mainly reflect those in the atmosphere (weather and climate). Water circulation is controlled by the mean wind patterns, tempered by effects due to the Earth's rotational and gravitational field such as the Coriolis effect and density differences. This results in four major circular currents or gyres (Fig. 74), clockwise in the North Pacific and North Atlantic and anticlockwise in the South Pacific and South Atlantic, which together with the Antarctic circumpolar current, dominate the surface flow pattern. In the Indian Ocean, the pattern changes seasonally because of the seasonal wind reversal of the monsoon, and in many other areas the currents show complex three-dimensional patterns on a smaller scale due to the continental configuration and to up- and downwelling. The position of the gyres dictates that off Peru and off the west coast of Africa surface water flows away from the continental margin and is replaced by colder water welling up from the deep-water mass (upwelling velocities up to 25 m/day). This compensates for the more diffuse downsinking of surface water as warm currents enter the polar regions, where the thermocline is weak. Water transport by most of the major currents lies between 15 and 50 million m³/sec: the

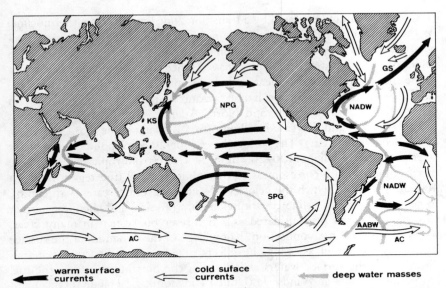

Fig. 74. Main circulatory patterns in the world's oceans showing the distribution of warm (black arrows) and cool (white arrows) surface water currents and the approximate movement directions of the deep water masses (grey arrows). Surface currents: GS, Gulf Stream; KS, Kiro Shio; AC, Antarctic Circumpolar; NPG, North Pacific Gyre; SPG, South Pacific Gyre. Deep water masses: NADW, North Atlantic Deep Water; AABW, Antarctic Bottom Water. (After Smith, 1982, and Turekian, 1976.)

largest transport figures are given by the Gulf Stream (150) and the Antarctic circumpolar (200) currents.

The circulation pattern of the deep-water mass (Fig. 74) is less well known and is mainly driven by density differences (thermohaline circulation). Direct measurement of currents at the seafloor normally indicates very slow movement (< 2 cm/sec), but near-bottom velocities of 15–20 cm/sec have been noted and current-generated bottom features including ripple and scour marks, sediment tails and deflected organisms indicating even higher flow rates are widespread. On a large scale, the global pattern is mainly deduced from the temperature–salinity distribution and the content of particulate matter at depth, using the concept of water masses. On the basis of this, volumes of water showing the same characteristics are interpreted as having undergone the same history. The main deep-water mass derived from the north is the NADW (North Atlantic deep water). This flows along the west side of the Atlantic until it penetrates and subdivides a deep-water mass derived from the south. These southern waters consist of the AABW (Antarctic bottom water), which is very cold but less saline and is derived from the Weddell Sea, and the AAIW (Antarctic intermediate water), which is derived from the South Pacific gyre and carried eastwards by the circumpolar current. The very saline NADW eventually turns eastwards into the Indian Ocean at a depth of 2500–3200 m and can still be identified in the Tasman Sea. The three-dimensional shape and precise dynamics of these partly diffusive, partly advective currents is still poorly understood, and the oceanographic models for deep-water circulation (on which, for instance, the International Atomic Energy Agency limits and restrictions for ocean dumping of radioactive waste are based) are probably oversimplified. All that can be said is that pollutants released into solution on the ocean floor at a carefully chosen site will take many hundred years (data from ^{14}C and other tracers indicate a residence time for water molecules in the deep-water mass of 1000–1500 years) to reach the surface and that this will take place at a location remote from the site, probably in another ocean.

Ocean chemistry is best thought of in terms of an upper and a lower reservoir, roughly corresponding to the surface layer and the deep-water mass, respectively. The main components of both are the cations Na^+, Ca^{2+}, Mg^{2+} and K^+ and the anions Cl^- and SO_4^{2-}. Two factors produce significant variations in some of these major ions and in many of the minor ones: the ocean currents described above and the biological activity. As the upper reservoir interacts with the atmosphere (through evaporation, precipitation, etc.), it receives influxes of chemicals from the rivers (denudation) and from volcanic activity (much of it submarine). Solar energy is converted into organic matter by the microflora (mainly phytoplankton), which in turn support complicated faunal and bacterial food chains. Organisms remove certain elements from the seawater, and when they die, their remains fall from the upper reservoir into the dark, cold, deep-water mass, where they nourish further bacteria and floating or bottom-dwelling organisms. The distribution of organisms in the upper reservoir is limited by certain critical elements such as P, Si and N (bio-limiting elements) and to a lesser extent by the concentrations of other elements such as Ca, C, Ba and Ra. These show a variable concentration, depending on the biological activity, and as the falling organic residues decompose, the biolimiting elements are released into solution and are carried along by the deep ocean currents, to be returned to the surface in areas of upwelling (recycling, with an average turnover of 1000 years) or to be transported from one ocean basin to another. For instance, the NADW becomes successively enriched in P, Si, and N on its way from the Atlantic through the Indian to the Pacific Ocean, as well as becoming richer in CO_2 (from decomposing organic matter) and correspondingly depleted in O_2.

III. DEEP-SEA SEDIMENTS

The aspects of marine geology (as opposed to oceanography) relevant to the radwaste disposal problem concern the composition, structure, and distribution of the sediments which cover the volcanic basement of the oceanic crust. Deep-sea sediments can be roughly subdivided into two groups according to the origin of the detritus and the mode of formation. Terrigenous sediments include all those derived directly from fluvial, glacial, or coastal erosion of the land masses around which they form a broad, irregular mantle (Fig. 75). They also include shallow-water shelf deposits of whatever origin (see Chapter 7, Section II,B), and they locally encroach on to the deep ocean floor through the action of sediment gravity flows (including slumps, debris flows, and turbidity currents) down the continental slopes. Pelagic sediments are formed far from the continental margins, where the main depositional process is settling out of the surface waters and drifting under the influence of the deep ocean currents. There are three main types of pelagic sediment,

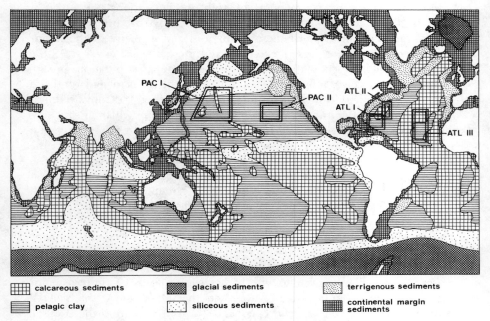

| calcareous sediments | glacial sediments | terrigenous sediments |
| pelagic clay | siliceous sediments | continental margin sediments |

Fig. 75. Distribution of the present-day sediments on the ocean floor, showing also the five study regions being evaluated by the Subseabed Disposal Program. Figure 76 shows an area located in study region PAC I, and Fig. 77 gives two sediment cores from that area (see also Fig. 93). (From Smith, 1982, by permission of Trewin Copplestone Books Ltd.)

named after the dominant mineral constituent: pelagic clay ($>60\%$ clay minerals and authigenic components), calcareous ooze ($<30\%$ clay, calcareous fossils dominant), and siliceous ooze ($<30\%$ clay, siliceous fossils dominant). Their distribution (Fig. 75) reflects variations in two main factors: the biological fertility of the surface water layer (determined partly by the water circulation pattern) (Fig. 74) and the water depth. Both types of ooze are of biogenic origin and accumulate at rates of ~ 1 cm/1000 yr, whereas pelagic clays are largely nonbiogenic (derived from wind-blown desert or volcanic dust and authigenic minerals) and accumulate much more slowly (~ 1 mm/1000 yr). Since $CaCO_3$ is the main skeletal material for organisms in the surface layer (of which by far the most common planktonic types are foraminifera, coccoliths, and pteropods), it forms the main component of the particulate matter sinking through the thermocline in areas of high biological fertility. However, $CaCO_3$ dissolves in seawater as soon as a certain degree of undersaturation is reached, usually at depths of 4000–5000 m. This calcite compensation depth (CCD), where the rate of supply of $CaCO_3$ is balanced by the rate of dissolution, is exceeded over large areas of the abyssal plains. If the surface fertility is high, only siliceous skeletons (remains of diatoms and radiolaria) are preserved below the CCD and siliceous oozes accumulate. Where surface fertility is low (e.g., in subtropical current convergence areas), the settling inorganic particles are not masked by the biogenic input and pelagic clays accumulate at a much slower rate. Since pelagic clay is one of the generic sediment types being considered as enclosing media for HLW disposal in the seabed, we now look at its properties in more detail, taking as an example the Northwest Pacific Basin.

Example: Pelagic Clays, Northwest Pacific Basin

Because of extensive deep-sea drilling, piston coring, sea-bottom photography, and echo-character mapping connected with the international Subseabed Disposal Program, the sedimentary deposits on the floor of the Northwest Pacific Basin are some of the best known deep-sea deposits (Fig. 76). On the abyssal plains surrounding the Shatsky Rise (water depths >5000 m), siliceous clays and oozes with interbedded volcanic ash layers dominate to the northwest, whereas brown pelagic clays (red clays) with occasional chert horizons are predominant to the southeast. Deep-sea cores show that this characterization applies to the upper 100–150 m of the total sedimentary column, suggesting that broadly similar conditions have pertained throughout the area for at least the last 20 m.y., possibly much longer. This pattern is broken by regions of more rugged and irregular seafloor relief (rises, ridges, seamount chains and isolated seamounts) with numerous outcrops of volcanic

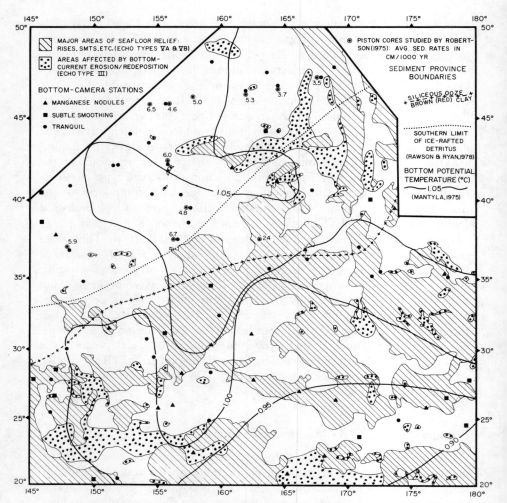

Fig. 76. Some results of echo-character mapping and other studies of the deep ocean floor in the neighbourhood of DSDP sites 576 and 578 (Shatsky Rise, northwest Pacific Basin, see Figs. 75 and 93). (From Damuth *et al.*, 1983, by permission of the Geological Society of America, Inc.)

basement and only a thin, discontinuous sedimentary cover. Associated with these features are indications of localized mass transport deposits (slumps and debris flows), and several areas show evidence of intermittent sediment re-working (erosion and redisposition), indicating that strong bottom current activity took place episodically at various times, although the present bottom water flow is very slow.

Deep-Sea Drilling Project cores from the same area (DSDP leg 86) give a more detailed picture of the sedimentary succession (Fig. 77). At site 576 (see Fig. 76), very homogeneous brown clays make up the top 55 m of the core and represent a time interval of ~ 60 m.y., giving a time-averaged sedimentation rate of only 0.5 mm/1000 years (assuming no erosional events). The homogeneity is enhanced by bioturbation, i.e., mixing of the material in the surface layers by burrowing organisms at the time of deposition. The brown colour indicates constant oxidising conditions whereby the small oxygen content of the deep water is sufficient to produce iron–manganese oxides and hydroxides because of the extremely low sedimentation rates. DSDP hole 578 was drilled in the pelagic clay–siliceous ooze transition west of the Shatsky Rise (see Fig. 76). The upper 76 m of the core consists of unoxidised (gray-coloured) siliceous clays and oozes, indicating predominantly reducing conditions during the past 2.5 m.y., although they are oxdising at the present time. During this time, sedimentation rates varied from 25 to 50 mm/1000 years, with the higher rates towards the top. The biogenic content of these sediments and the relatively high sedimentation rates reflect Pleistocene oceanographic conditions (surface current pattern and bottom current activity). These must have been markedly different to earlier conditions in the same area, since below 76 m the core shows dark brown pelagic clays, similar to those of site 576, and with an average sedimentation rate of only 1 mm/1000 years. The mineralogy of the brown clays is highly variable in detail (variable amounts of volcanic ash, eolian quartz and authigenic zeolites). The main clay mineral is illite (35–45% by weight), with kaolinite next (5–15%) and minor amounts of smectite and chlorite. In general, all the work reported from this area indicates that the deep-sea sedimentary environment is not as uniform and monotonous as it has often been portrayed to be.

IV. RELEASE OF RADIOACTIVITY INTO THE OCEANS

Ocean water has a natural radioactivity level of about 320×10^{-12} Ci/litre. This is due to its content of primordial, long-lived radionuclides (e.g., ^{40}K, ^{87}Rb, ^{235}U, ^{238}U and ^{232}Th) and their short-lived daughter products, and to the incorporation of radioisotopes of relatively short half-lives produced by cosmic interaction with atmospheric constituents or extraterrestrial materials (e.g., ^{3}H, ^{7}Be, ^{10}Be, ^{14}C, ^{26}Al and ^{32}Si). In addition, a small proportion of the ocean inventory (~ 0.1%) is due to artificially produced radionuclides from surface nuclear explosions and from accidental or planned releases by nuclear power stations, reprocessing plants, and transport systems (satellite reentry, aircraft and submarine loss, etc.). Since the signing of the Treaty

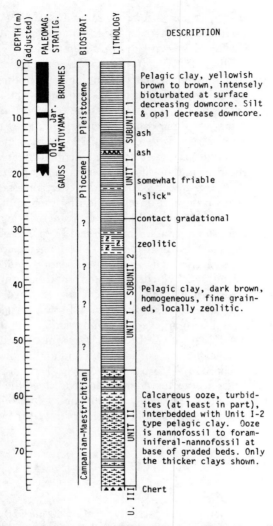

Fig. 77. Summary of preliminary drilling results from Deep-Sea Drilling Program sites 576 and 578 in the northwest Pacific Basin (see Fig. 76). These sites lie in candidate locations E_2 and B_1 of the Subseabed Disposal Program, which is evaluating the feasibility of disposing HLW in the deep ocean sediments (Fig. 93). (From Shephard, 1983.)

Banning Nuclear Weapons Testing (1963) and since the shutdown of the reactors on the Hanford Reservation, the United States, which used Columbia River water as the primary coolant in a flow-through system (see Chapter 7, Section II,A), the only intentional releases of radioactive materials to the

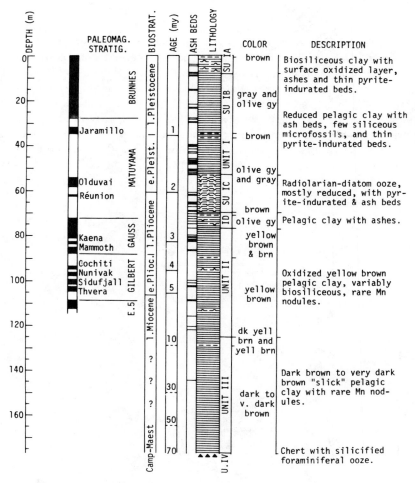

Fig. 77. Continued.

ocean waters today are the liquid effluents from British, French, and Indian nuclear fuel reprocessing plants and, to a lesser extent, those from atomic-powered merchant and military vessels. In 1974, this amounted to 1200 Ci of α-activity (equivalent to 20 kg of Pu) from Windscale alone. In addition, the dumping of radioactive wastes in the ocean, an activity which has now been halted (see Chapter 2, Section III,A), may result in slight releases of radionuclides to the ocean water in the future.

Although global comparisons indicate that these emissions are negligible in amount, natural concentration processes result in local radioactivity levels which are a cause of increasing concern. In addition to the effects of complex ocean water flow patterns (Fig. 78), sediments and organisms absorb and accumulate radionuclides preferentially. The best known example of this is the red alga *Porphyra*, which is used in South Wales for making laverbread and whose radioactivity increases from a normal level of 5000–15,000 pCi/kg to as much as 375,000 pCi/kg in the coastal waters around the Windscale plant. As in the case of radionuclide migration from underground repositories (see Chapter 11), the general tendency of the relevant natural processes is clearly towards dilution and dispersal, but this may be reversed locally by site-specific concordance of poorly understood but critical factors.

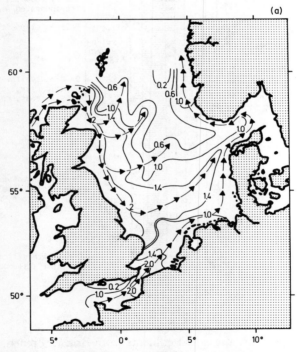

Fig. 78. Distribution of radionuclides introduced by humans into the ocean waters. (a) Water movement in the North Sea deduced from the concentration of ^{137}Cs (in picocuries/litre) in the seawater 4 years after major releases of this radionuclide from the reprocessing plants at Windscale (United Kingdom) and La Hague (France). (b) Distribution of tritium (^{3}H) in the Atlantic Ocean, showing the effect of the NADW (see Fig. 74) in carrying it to great depth. The tritium originated as fallout from nuclear bomb testing about 10 years before these measurements were made. (From Gerlach, 1981, by permission of Springer-Verlag.)

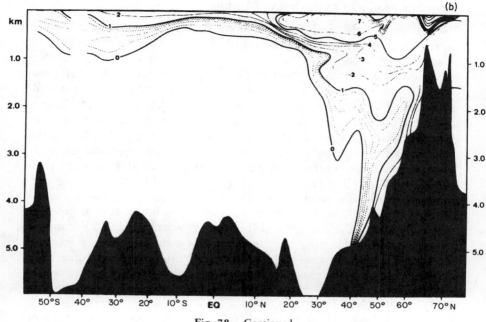

Fig. 78.   Continued.

The experience gained during and after such full-scale tracer experiments (Fig. 78) has and will prove invaluable in assessing the oceanographic aspects of future releases which may take place from solid and packaged radwaste which have been or may be placed on or in the ocean floor. They have already shown that the spatial and temporal variability of fluctuations in the mean flows are still essentially unpredictable. Since the seas are, however, obviously well mixed on a time scale of thousands of years, the aim of ocean disposal is containment until mixing at this rate could not result in harmful concentrations. In the case of LLW dumping activities (see Chapter 2), release of radionuclides to the deep ocean water depends only on the integrity of the container (which may be endangered by pressure cracking, interaction with bottom fauna or corrosion processes) and the resistance of the waste matrix to dissolution (a function of its chemical stability, effects of biological activity and bottom water circulation, etc.). For the proposed subseabed emplacement of high-level wastes, a more efficient sequence of barriers is envisaged, of which the main one is the sedimentary enclosing medium, exemplified by the following model.

Deep-Sea Sediment as Enclosing Medium

Work on the subseabed disposal concept is being supported by land-based experimentation and release pathway analysis similar to that already discussed in some detail in connection with land disposal concepts (see, e.g., Chapter 7, Section IV,A; Chapter 8, Section IV; Chapter 10, Section V). After successful emplacement (i.e., undamaged canister at 30–100 m depth in a well-characterized sedimentary environment with complete closure of the emplacement hole), the only possible natural release mechanism is thought to be corrosion of the waste canister and upward migration or diffusion of the radionuclides through the sediment column to the sediment–water interface. It is proposed to design the canister for an endurance of 500 years in a thermal field in which the canister wall never exceeds 250°C (by adjusting canister material and dimensions, waste loading, and storage time). The maximum extent of the 100°C isotherm in the surrounding sediment is then expected to be < 1 m from the canister wall (Fig. 79). Water trapped in the sediment in this near field will become less dense than that in the surroundings (at these pressures, a density increase of 15% occurs on raising the temperature from 2 to 250°C) and will tend to migrate upwards in a convective system. However, the very low permeability of deep-sea sediments (for pelagic clays, hydraulic conductivities are 10^{-6}–10^{-7} cm/sec) will ensure that such movement will not exceed 30 cm in 1000 years (the approximate length of the thermal period) and that most of the heat is dissipated by conduction. Even if thermally induced water movement were to be more important, the canister would prevent any contact between the sediment–water system and the waste matrix until after the main thermal pulse had subsided. This will also be necessary to hinder a significant thermochemical effect: above 200°C seawater–sediment mixtures become distinctly acidic and many trace elements become soluble. In the absence of a thermal perturbation, existing data suggest that natural pore-water movement is controlled by compaction and is of the same order of magnitude as the local rate of sedimentation (0.1–10 mm/ 1000 years). Thus, in the absence of a natural geothermal anomaly, pore water from deep in the sediments will never reach the sediment–ocean water interface. In this environment, the main mechanism of radionuclide migration through the water-saturated sediments after the canister is breached is by molecular diffusion, a very slow process even more retarded by the high sorptive capacity of deep-sea clays (Fig. 79). Apart from very unlikely catastrophic events (e.g., onset of volcanism or faulting or human intrusion during deep-sea mining activities), the only factor which could cause increased diffusion rates would be the formation of complex anions (e.g., complexation of transuranics with carbonate ions), since the anion exchange capacity of highly oxidised sediments is known to be negligible. However, even without

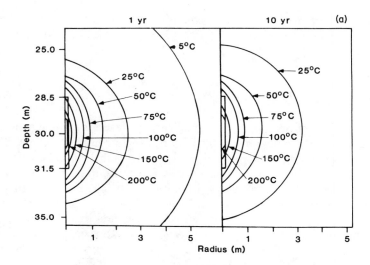

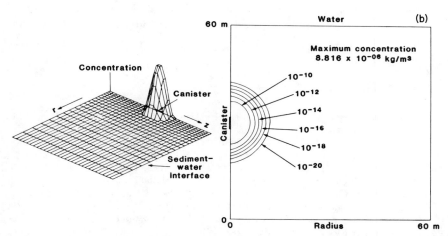

Fig. 79. Examples of model calculations to estimate effect of emplacing HLW canisters at a depth of 30 m in deep-sea clays below the ocean floor. (a) Isotherms around a HLW canister 1 and 10 years after emplacement, assuming an initial heat production due to radioactive decay of 1.5 W. The 100°C isotherm, within which significant hydrothermal alteration of the sediment is expected, never extends more than 80 cm from the canister. (b) Distribution of ^{239}Pu in the sediment around a HLW canister 100,000 years after emplacement, assuming instantaneous breaching. Because of the high sorption of ^{239}Pu by the surrounding clays, virtually none will ever reach the water–sediment interface. (From Hollister *et al.*, 1981, by permission of the American Association for the Advancement of Science.)

sorption, conservative diffusion coefficients give breakthrough times to the sediment–ocean water interface of 10^5–10^6 years, depending on radionuclide and depth of emplacement.

V. CONCLUDING REMARKS

Although recent oceanographic investigations, seafloor mapping, and deep-sea drilling and sampling have revealed surprisingly complex regional fluctuations super-imposed on the relatively well-known global patterns, strictly controlled ocean disposal still remains a viable alternative to land-based concepts, from the scientific and technological point of view. The main advantages are that past and present conditions can be defined on the basis of random sampling (uniformity on a 10–100 km distance scale) and that future conditions have a very high probability of not undergoing drastic changes (predictability on a million-year time scale). On the other hand, uncontrolled disposal of radioactive materials in surface waters or shallow shelf seas is now recognized as bad practice, since it may result in unpredictable local accumulations of harmful amounts of radionuclides due to poorly understood, localized concentration processes.

SELECTED LITERATURE

Oceanography and marine geology: general reviews: Turekian, 1976; Harries, 1980; Gerlach, 1981; International Atomic Energy Agency (IAEA), 1981, 1983d; Kennett, 1982; Murray and Stanners, 1982 (pp. 255–271); Smith, 1982 (pp. 311–324); Committee on Marine Geo-sciences, 1983; Siever *et al.*, 1983 (pp. 100–112); Shackleton *et al.*, 1984.
Release of radioactivity to the marine environment: IAEA, 1978b, 1983d; Gomez *et al.*, 1980; Bowen and Hollister, 1981; Murray, 1981; Templeton and Preston, 1982; Organization of Economic Cooperation and Development (OECD), 1983a; Sholkovitz, 1983.
Properties of deep-sea sediments as related to radwaste disposal: Dawson, 1978; Neiheisel, 1979; OECD, 1980b (pp. 179–253); McVey *et al.*, 1979; Chavez and Dawson, 1981; Hollister *et al.*, 1981; McCall and Tevesz, 1982; Murray and Stanners, 1982 (pp. 240–254); Damuth *et al.*, 1983; Shephard, 1983 (pp. 17–172); Subseabed Disposal Program, 1983.

CHAPTER 13

Climatic Change
and Continental Glaciation

I. INTRODUCTION

The rise of oceanography in postwar years not only contributed to a fundamental revolution in the earth sciences (see Chapter 3) and allowed a serious assessment of the limits and possibilities of exploiting the ocean floor (see Chapter 12), it also made possible a complete reevaluation of climatic history, particularly for the last 2–3 m.y. The continuous record preserved in the ocean sediments confirmed and redefined what had been known from the fragmentary land record for over a century, i.e., that the Pleistocene was a period of strong and rapid climatic change which resulted in large fluctuations in the polar ice masses. Since the period of these fluctuations (10^3–10^5 years) lies within the nominal isolation time for HLW (see Chapter 1, Section IV), and because there is every indication that we now stand between two glacial periods, predicting future climatic change has become an important aspect of performance assessment for radwaste repositories. In this chapter, we look first at the evidence which suggests a renewed development of continental ice sheets, i.e., at the Pleistocene climatic history and at the indications that this follows an astronomically controlled cyclicity which can be projected into the future. Then we consider the problem of reconstructing the size, shape and dynamic state of the last great ice sheet in the northern hemisphere, which reached its maximum extent about 20,000 years ago. In a final section, such an ice sheet is considered as a worst case for future events and an attempt made to assess the effects it might have on radwaste repositories. As with all the other chapters in Part II, emphasis here is on understanding the processes involved. The integration of this data into a general performance assessment based on analysis of the whole repository system is the subject of Chapter 14.

II. PLEISTOCENE CLIMATIC CHANGE

The Great Ice Age has been a well-established part of public knowledge since the 1840s, when the exotic blocks and the blankets of "drift" throughout Europe were recognized to be of glacial origin by Alpine naturalists like Louis Agassiz. The Flood was finally abandoned as a geologic agent, and the erratics, moraines, varved sediments, loess deposits, and the myriad of geomorphological features allowed the reconstruction of an elegant picture of advancing and retreating continental ice sheets, of which the Antarctic and Greenland ice caps are the meagre present-day remnants. By the 1950s, an enormous body of data had been amassed concerning this latest great event of Earth history, spurred by the importance of the Pleistocene for understanding the evolution of human beings (see Fig. 33) and of Pleistocene deposits for the growth of civilization (agriculture, water supplies, foundation problems, building materials, etc.). Although it was recognized that each successive glacial advance tended to destroy the evidence of earlier episodes, the land record in the northern hemisphere seemed to indicate four major periods of world-wide cooling (glacials) separated by warm periods (interglacials), during the last 2 m.y.

This was the position before oceanographers started pulling cores from the sediments on the ocean floor which indicated a more complicated pattern of events. Critical parameters showed continuous variations forming complex patterns which, because the sediments do not show erosional breaks, are now considered as better indicators of climatic history (as opposed to the climate at particular localities and points in time) than the incomplete records pieced together on land. In the following, therefore, we reverse the historical development and look first at the pattern of climatic change as deduced from ocean sediments, before considering the land record of the last glaciation and the possible causes of glacial–interglacial periodicity.

A. The Ocean Record

The revolution in Pleistocene studies initiated by the study of deep-sea cores culminated in the pulling together of the various strands of evidence in the CLIMAP (Climate: Long-range Investigation Mapping and Prediction) project, instituted in 1971. This was an interdisciplinary, and later, international research programme aimed at integrating paleomagnetic, stratigraphic, isotopic, and oceanographic data to map the surface of the Earth during the last Ice Age and to measure the oscillations of Pleistocene climate. The ocean record had already revealed large variations in surface water temperature (e.g., from fluctuations in the abundance of certain microfossils and in the oxygen isotope ratios in calcareous microfossils) and a complicated sequence of sea level changes (e.g., from dated terraces and former shorelines), as expected from the evidence on land. Also, some progress had been made in dating these variations using an increasingly refined biostratigraphy coupled with newly developed techniques of radiometric and paleomagnetic dating, but a generally applicable correlation scheme and an integrated synthesis had been lacking. CLIMAP set out to develop a stratigraphic scheme which could be applied to every ocean and found that the key to this lay in the oxygen isotope ratios (Fig. 80).

The oxygen isotope method measures the ratio of ^{18}O to ^{16}O in the sample, usually a fossil foraminifera, and compares it to a standard, generally the fossil

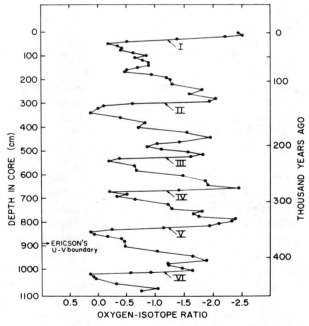

Fig. 80. Climatic variations over the last 400,000 years as recorded by changes in the oxygen-isotope ratios in a deep-sea core V12–122 from the Caribbean. The main maxima correspond to the maximum extent of the continental ice sheets, each preceded by a slow, fluctuating buildup and terminated by a rapid deglaciation event (I–VI). (From Imbrie and Imbrie, 1979, by permission of the authors.)

Belemnitella americana from the Pee Dee Formation (the PDB standard). Deviations from the standard (δ) occur because the two isotopes react slightly differently to physicochemical processes such as evaporation, condensation, and crystallization, and the sample ratio reflects the H_2O/CO_2 environment during the organism's lifetime, as the oxygen was being incorporated into the $CaCO_3$ of its shell. Although the oxygen isotope ratio of ocean water is affected by many factors, it is now generally recognized that the main factor over the past million years has been the mass of water in the polar ice caps. Development of a large ice mass results in an enrichment of ^{18}O in the oceans, because the combined effects of evaporation and condensation results in glacial ice considerably enriched in ^{16}O. Continuous profiles of δ in ocean sediment cores (Fig. 80) show a distinctive sawtooth pattern, with δ rising slowly and with many minor oscillations to major peaks (maximum ice mass = glacial) spaced approximately 100,000 years apart, followed by rapid drops in δ to major troughs (minimum ice mass = interglacial). The last peak occurred about 20,000 years ago, corresponding to the last glacial maximum known on land.

B. The Land Record

During the same period that the ocean floor was proving such a faithful recorder of world climatic conditions, land-based studies were also revealing an ever more complicated situation, which could only be reconciled with the classical four-glacial picture with difficulty. The main impulses came from studies outside the areas of moraines, glaciofluvial deposits and terraces which mark the successive positions of the ice sheet margins, i.e., from the loess–paleosol successions representing alternating periglacial dessert conditions (during a glacial maximum) and forest cover (during an interglacial), and from the continuous sedimentary records in existing or drained Pleistocene lakes. Although the land record of the earlier Pleistocene glaciations will always be incomplete, the history of the last 20,000 years is now known in considerable detail (Fig. 81) and is being continually refined using a whole spectrum of dating techniques (including varve chronology, tree-ring dating, a range of isotopic techniques, paleomagnetic loess stratigraphy, amino acid racemization and thermoluminescence). This is being complemented by research in other fields, such as glaciological studies on ice cores from, and on exposed old ice in, Antarctica and Greenland and archeological studies.

C. Milankovitch Cycles and Future Climate

The continuous record of cyclic climatic change preserved in the deep-sea sediment cores (Fig. 80) can be correlated with systematic changes in the earth's rotation axis and in the shape of its orbit. Orbital variations had been

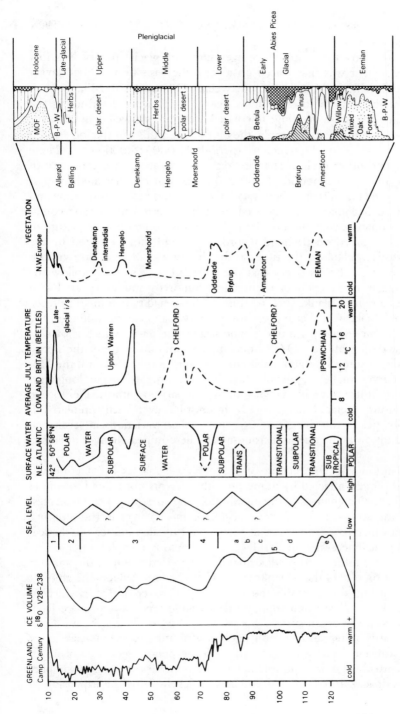

Fig. 81. Correlation of the ocean record of climatic changes over the last 120,000 years with the European land record, as preserved in the glacial deposits (fossil beetles and plants). (From Bowen, 1978, by permission of Pergamon Press.)

considered as a possible cause of ice ages since the glacial theory was first propounded, but it was first placed on a scientific basis by the Yugoslavian mathematician Milutin Milankovitch between 1920 and 1930. Milankovitch showed that the amount of summer radiation (the factor determining the amount of ice and snow melting) fluctuated according to well-defined astronomical laws governing, for example, the tilt of the Earth's axis (period 44,000 years) and the precessional movements (periods 22,000–26,000 years). These fluctuations and other cyclical changes (e.g., the variation in eccentricity of the Earth's orbit over a period of 100,000 years) can be calculated for past and future time (Fig. 82) and show regularly recurring radiation maxima and minima which Milankovitch suggested must be reflected in climatic conditions. Unfortunately, they did not fit at all with the contemporary picture of four glaciations based on the land record, nor could they be tested by the dating methods available at the time, so the theory was shelved, not without some controversy. The discovery that the $^{18}O/^{16}O$ curves closely follow the calculated radiation curves over the last few hundred thousand years led in the 1970s to a rapid Milankovitch revival which has had two dramatic effects. First, spectral analysis of the oxygen isotope curves and careful tuning to the Milankovitch curves has permitted a time scale to be set up which allows an independent dating of the $^{18}O/^{16}O$ maxima and minima without the necessity of making assumptions about the rate of sedimentation. Second, the Milankovitch curves make it possible for the first time to make long-term predictions of future climatic change and, by comparison with the Pleistocene record, of future glacial events (Fig. 82). Interest has even been stimulated in other possibilities in this direction, e.g., that cyclical variations in seismic or volcanic activity may be due to astronomically determined changes in crustal stresses.

The expansion and contraction of the polar ice masses is, of course, a complicated process affected by numerous other factors not related to orbital variations, including changes in the distribution and size of land and sea areas, changes in the composition of the atmosphere and changes in the intensity of the Earth's magnetic field. Even in the absence of such influences, the ice, ocean and atmosphere system is subject to numerous complex interactions whose effects are difficult to predict quantitatively. Reduction in the amount of solar radiation due to the Milankovitch effect not only reduces the amount of ice melting but also increases the area of snow cover on land, which in turn increases the albedo and supports the cooling tendency (positive feedback). However, the ocean surface waters, at first relatively warm and therefore with relatively high evaporation rates (contributing to the increase in ice volume by causing increased precipitation in the polar regions), cool down rapidly, eventually starving the ice caps of the necessary new snow supplies (negative feedback). In contrast, decreasing sea level during the expansion of

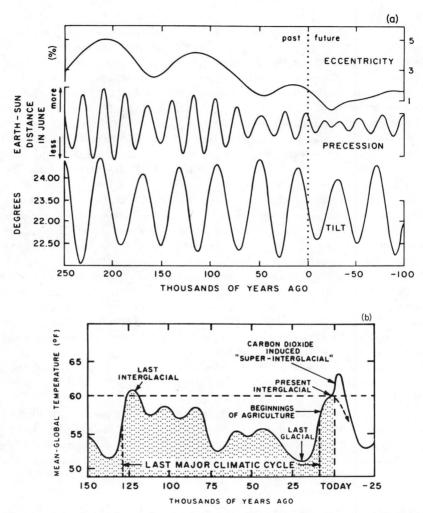

Fig. 82. Predicting future climatic change. (a) Changes in the geometry of the Earth's orbit (eccentricity, precession, tilt) cause variations in the pattern of incoming solar radiation (the Milankovitch effect). These changes can be calculated for past and future time. (b) Synthesis of the last 150,000 years of climatic change in terms of mean global temperature and a projection into the future, based on the Milankovitch effect and an estimation of the changes in CO_2 content of the atmosphere (due to deforestation and the burning of fossil fuels). (From Imbrie and Imbrie, 1979, by permission of the authors.)

the ice sheets decreases the area of very shallow water on the continental shelves, thus reducing the number of $CaCO_3$-building organisms. This effectively reduces the amount of CO_2 in the atmosphere, causing the reverse of the greenhouse effect, i.e., further cooling (a second positive feedback). When the amount of solar radiation increases due to the Milankovitch effect, similar feedback loops work in reverse, but apparently in different combinations and at different rates: deglaciation seems to take place much more rapidly and completely than ice buildup, giving the typical asymmetry of the oxygen isotope curves (Fig. 80). In some areas there is evidence that most of the ice melting during the last glacial retreat took place between 11,000 and 6000 years BP, causing a world-wide rise in sea level of 100 m in a mere few thousand years.

In addition to these effects, there is the suspicion that the Arctic and Antarctic ice sheets may not be in phase. Some evidence indicates that the Antarctic ice mass may actually continue increasing in size during a period of increasing solar radiation (the start of an interglacial in the Northern Hemisphere), due to increased evaporation and precipitation from the circum-Antarctic surface waters. Once a critical size is overstepped, when most of the basal ice reaches the pressure melting point, very rapid expansion may take place, facilitated by shear heating in the more rapidly flowing ice (a "surge" on a time scale of 100–1000 years). This is expected to form a large ice shelf, significantly increasing the Earth's albedo and counteracting the global warming due to the Milankovitch effect. Quite apart from its effect on global climate, however, the possibility of a surge of this type places a question mark against the idea of using the Antarctic ice sheet for radwaste disposal (see Chapter 2, Section IV,E).

In spite of these uncertainties, the most generally accepted climatic forecast indicates the occurrence of a new glacial period about 25,000 years hence, even though the warming effect of increasing the CO_2 content of the atmosphere by the burning of fossil fuels and by deforestation is likely to dominate global climate over the next 2000 years (Fig. 82).

III. RECONSTRUCTING THE LAST GREAT ICE SHEET

The best known glacial event is the one called the Wisconsin, Weichselian, or Würm glaciation in the Northern Hemisphere, which reached its maximum extent about 20,000 years ago and is related to the final peak on the $^{18}O/^{16}O$ curve. Since this volume of ice seems to be representative of the ice masses developed at various times throughout the Pleistocene, it can be taken as a worst case for the future. Therefore, attempts to reconstruct the dimensions and dynamics of this ice maximum are the first step towards predicting the effects of future ice ages on repositories. The results of one such reconstruction is shown in Fig. 83. Although such reconstructions represent a synthesis and interpretation of many lines of evidence, often still controversial, they are mainly based on the results of studies in glacial geology and glaciological–geodynamical modelling.

A. Glacial Geology

Glacial geology provides the basic input data and boundary conditions for the reconstruction models. Based on the detailed surveying of the multifarious traces of ice extent and movement (such as erratics, moraines, terraces, and striations), it contributes three main types of evidence. First, it identifies the positions of the ice margins on land and attempts to date ice margin fluctuations. There is an enormous volume of literature on this topic, and the irregular retreat of the Wisconsin or Weichselian ice margins over the last 20,000 years in many areas is known in great detail (e.g., Fig. 44). An important finding of this type of work has been that major readvances over distances of hundreds of kilometres have taken place on a time scale of 1000 years and are not recorded in the ocean sediments, where ice volume variations are only distinguishable on a scale of 10,000 years. An important limitation, however, is that the ice margin in oceanic areas remains unknown and introduces a high degree of uncertainty into the models. Second, glacial geology attempts to determine the ice flow directions. This is more controversial, not because there is any lack of directional data (striations, flutes, drumlins, provenance data, etc.), but because it is difficult to date and because there is evidence that local flow regimes may diverge strongly from the regional pattern and may change significantly with time. Third, Pleistocene geology in its widest sense provides the evidence for crustal movements related to glacial loading (marginal lakes, tilted shoreline terraces and neotectonics) and details worldwide sea level changes related to ice buildup or melting.

B. Glaciological–Geodynamical Modelling

Most attempts to model Pleistocene ice sheets are based implicitly or explicitly on the assumption that their physical state (temperature distribution, flow characteristics, etc.) was similar to that of the Antarctic and Greenland ice sheets of today. From earlier two-dimensional flow-line models, emphasis has now changed to three-dimensional modelling similar to that of the CLIMAP project. The basic input data into such models includes the topography of the ice bed, net mass balance information, ice sheet response data, estimates

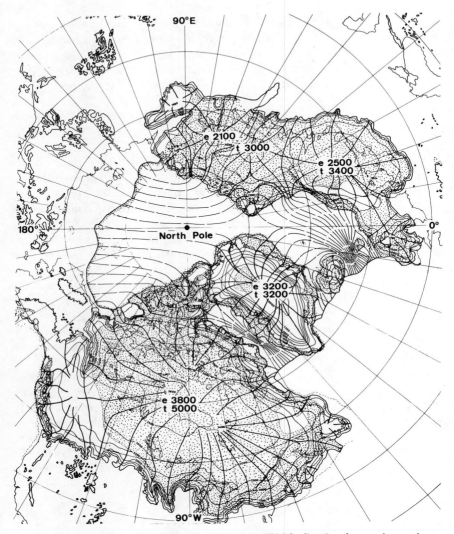

Fig. 83. A reconstruction of the late Wisconsin–Weichselian ice sheet at its maximum extent 20,000 years ago, showing regional flow lines and the estimated elevation (e) and thickness (t) of the main centres. Stippling indicates areas where the base of the ice was below contemporary sea level (From Denton and Hughes, 1981, by permission of John Wiley and Sons, Inc.)

of isostatic readjustment and Milankovitch radiation variations, in addition to the size and flow-line determinations derived from glacial geology. Whether static or dynamic, however, such models are based on glaciological estimates of such parameters as total ice volume, thickness variations and basal temperatures. Today these are being refined by applying glacioisostatic data. Previously, the volume of an ice sheet was estimated independently (e.g., from sea level changes) and the isostatic response used to determine the thickness and mechanical state of the lithosphere on the 10^4-year time scale. As the rheology of the Earth's interior is becoming better known from other kinds of data, the problem can now be inverted: an Earth model can be used to reconstruct ice sheet dimensions. Combined glaciological–geodynamical modelling has now reached a level of sophistication which allows the future dynamics of the Antarctic ice sheet to be predicted with different mass balance scenarios.

IV. CONTINENTAL GLACIATION AS A DISRUPTIVE EVENT

The advent of a new glacial, comparable in size to the one described above, will be a disruptive event of first magnitude. From the past record, it is sure to occur at least once in the next 100,000 years and it will probably take place about 25,000 years hence. Future civilizations will have to be able to accommodate themselves to advancing ice margins, cooler and wetter climates, subsiding sealevels, and changing vegetation and ocean circulation patterns. Today's LLW repositories will by that time be forgotten, since they will have long since ceased to pose a greater risk to health than other earthworks. The chances are also high that concern about the HLW repositories inherited by these civilizations from our Atomic Age will be swamped by more serious and immediate preoccupations. It may seem an academic and pointless occupation to speculate today about the effect of a future glaciation on a HLW repository many tens of millenia in the future. It may, in fact, be so. Nevertheless, as we shall see in Part III, proof of the safety of radwaste disposal must be based on realistic modelling of future processes and events. This depends to a large extent on the quality of the input data, and some types of input data, such as those defining the specific geological situation of the repository, are obviously strongly affected by climatic change and continental glaciation. Out of this problematic area we take, in the following, two aspects (glacial erosion and disturbance of the hydrogeological regime) as examples of kinds of effects to be expected.

A. Erosional Effects

The erosional effects of a continental glaciation include glacial scouring, glaciofluvial erosion, increased denudation in nonglacial regions (due to increased rainfall, desertification, changing coastlines, etc.) and the effects of local, catastrophic events, such as the bursting of ice-dammed or moraine-dammed lakes. The maximum influence of many of these processes can be estimated from present-day observations and from interpretations of the Pleistocene land record. Perhaps most uncertainty surrounds the question of glacial scouring below an ice sheet, and this is considered in more detail in the following.

Glacial scouring is produced by a combination of abrasion (scraping of debris-laden ice across its bed) and quarrying (plucking of fragments out of overridden bedrock). The relative importance of these two processes vary according to bedrock geology (including preglaciation topography, weathering, permeability and jointing) and ice flow regime, but the total scour effect depends on whether or not the ice is frozen to its bed, which is in turn dependent mainly on ice thickness, ice surface temperatures and ice velocity (shear heating). When basal ice temperatures are below the pressure melting point over large areas, landscapes of little or no scouring remain when the ice retreats (Fig. 84) and in some areas covered by the Wisconsin or Weichselian ice sheet even the preglacial regolith has been preserved. This situation is most likely to pertain below thin ice in a cold continental climatic regime. When the basal ice is everywhere at the pressure melting point, scouring is everywhere active, and the resultant landscapes are typically glaciated over wide areas. Estimates of the average thickness of rock removed by the Wisconsin or Weichselian ice sheet from landscapes of areal scouring vary from a few metres to a few tens of metres, but occasionally much higher figures are cited. A third type of landscape shows selective linear erosion in which deep troughs are excavated in a plateau on which little or no scouring has taken place. These troughs, representing preglacial depressions in which the basal ice attained pressure melting temperatures and through which therefore much of the ice drained, reach depths of many hundreds to a few thousand metres, often far below sea level (fjords). The incidence of this type of subice erosion would obviously be a direct danger for most HLW repositories. It is thought that areal scouring is favoured by a maritime climatic regime, since this is often associated with high surface temperatures and high ice velocities, whereas selective linear erosion is favoured in preglacial upland regions.

B. Disturbance of the Hydrogeological Regime

Apart from the possibility of exhumation or encroachment by glacial erosion and related effects, the main disruptive action of a new continental glaciation and associated changes in world climate will be to change the hydrogeological regime. This will take place by the complex interaction of different processes, making the end result extremely difficult to predict. At a particular site, a succession of new glacial and interglacial periods during the life of a proposed repository will cause alterations in some or all of three main parameters.

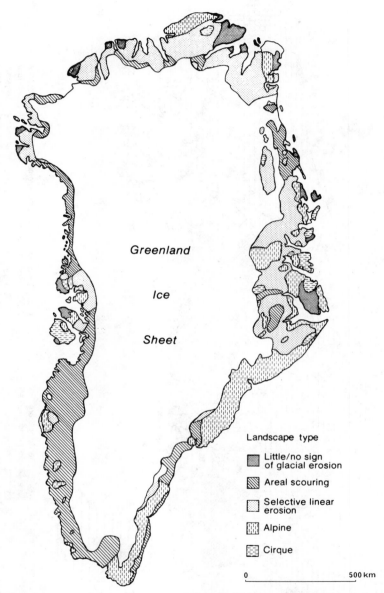

Fig. 84. Glacial landscape types around the present-day Greenland ice sheet, showing the different types and intensity of glacial erosion caused by continental glaciation. (From Sugden and John, 1976, by permission of the authors.)

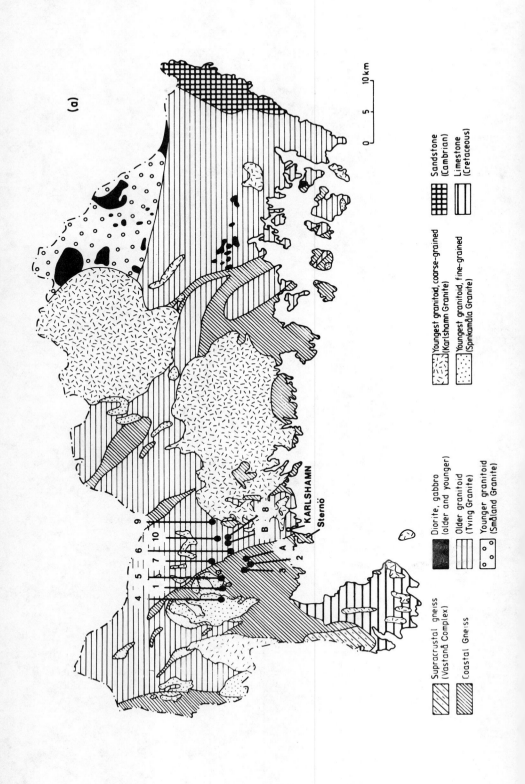

(a)

Supracrustal gneiss
(Vástaná Complex)

Coastal Gneiss

Diorite, gabbro
(older and younger)

Older granitoid
(Tving Granite)

Younger granitoid
(Småland Granite)

Youngest granitoid, coarse-grained
(Karlshamn Granite)

Youngest granitoid, fine-grained
(Spinkamåla Granite)

Sandstone
(Cambrian)

Limestone
(Cretaceous)

KARLSHAMN

Sternö

0   5   10 km

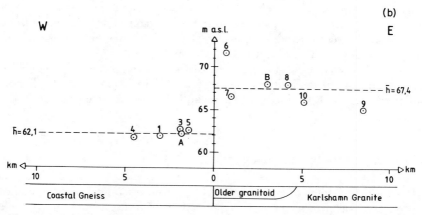

Fig. 85. Example of fault movements possibly associated with the retreat of the Scandinavian ice sheet about 10,000 years ago. (a) Geological map of the Bleklinge area, southern Sweden. Sternö is one of the localities being investigated with respect to the siting of an underground HLW repository (see Fig. 94). (b) Careful surveying of the boundary between washed and nonwashed till in the glacial deposits, combined with precise dating, shows that the contact between the Coastal Gneiss of Sternö and the Karlshamn Granite was a zone of differential movement (displacement 5–6 m) about 10,900 years ago. (From Björkman and Trägardh, 1982, by permission of the Geological Society of Sweden.)

Groundwater Flow Pattern. The erosive effects of ice, meltwater and lake bursts can significantly alter the surface drainage patterns, which, together with changes in the quantities of surface water, alter the amount and location of recharge. Hydraulic gradients and flow directions are also affected by the crustal response to ice loading, by the masking effect of the ice sheets, and by sea level changes.

Permeability Relations. Glaciation strongly affects the hydraulic properties of various materials in a variety of different ways. Permafrost reduces the hydraulic conductivity of aquifers by several orders of magnitude and its configuration significantly affects the subsurface flow patterns. The crustal response of bedrock to ice loading and unloading (flexuring and high subglacial fluid pressures) promotes renewed movement on old fault and joint planes (Fig. 85) and, in some areas, the formation of new fracture systems, thus affecting the bulk permeability and the water pathways. Many glacial deposits are extremely fine-grained and clay-rich, providing new barriers to percolation, and deposits of many types become more compacted and impermeable when overridden by ice or submerged and buried along coastlines.

Groundwater Chemistry. Climatic changes affect weathering and the dissolution of minerals in the subsurface by changing the flow rates and by changing the water chemistry. For instance, higher precipitation and cooler climate increase the CO_2 content of rainwater, increasing thereby its capacity for dissolving carbonates. Also, rising or falling water tables may affect such features as the redox conditions in a repository, or presence or absence of saline groundwaters (see Chapter 11, Section II,C).

On the basis of the geologic record of Pleistocene and earlier epochs, it may be possible to define the range of probable variation in some of these parameters at a specific site and to make conservative estimates of future release rates.

V. CONCLUDING REMARKS

In this final chapter of Part II, we have considered some aspects of one of the few areas of earth science where predictions of future events can be made with some confidence. At least over the last million years, extreme changes of world climate have been documented, which closely follow astronomically determined variations in solar radiation. Recent recalculations of these effects, coupled with detailed study of ocean sediment cores, have established a climatic history which can be extrapolated 25,000 years into the future and which indicates the recurrence of an ice age similar to the continental glaciation which reached its maximum 20,000 years ago. For a particular HLW repository site, this means that present-day conditions will certainly change dramatically during the early part of its life and that performance assessment must take account of possible changes. This is particularly true of the hydrogeological regime, which is one of the main factors affecting the performance of any deep-mined repository.

SELECTED LITERATURE

Ice Ages and future climatic change: Emiliani, 1966; Kukla, 1977; Bowen, 1978; Imbrie and Imbrie, 1979; International Atomic Energy Agency (IAEA), 1980b (pp. 385–406); Geological Society of Stockholm, 1982; Revelle, 1982; Schubert and Yuen, 1982; Beget, 1983; Nilsson, 1983; Deep Sea Drilling Project, 1983; Robin, 1983; Siever et al., 1983 (pp. 114–130); Blatter et al., 1984; Shackleton et al., 1984.
Reconstructing continental ice sheets: Mahaffy, 1976; Vorren, 1979; Clark, 1980; Budd and Smith, 1981; Denton and Hughes, 1981; Peltier, 1981; Andrews, 1982.
Continental glaciation as a disruptive event: Sugden and John, 1976 (pp. 151–212); Mörner, 1977; Potter, 1978; Freeze and Cherry, 1979 (pp. 163–165, 517–518; Davis, 1980; Mörner, 1980; Lawrence Livermore Laboratory, 1980a,c; Organization of Economic Cooperation and Development (OECD), 1981b (pp. 111–129); Björkman and Trädgårdh, 1982; IAEA, 1982b (pp. 728–736); Merrett and Gillespie, 1983; Swedish Nuclear Fuel Supply Co., 1983 (pp. 8:1–8:49).

PART III

APPLICATION

In Part II, we have been concerned with individual geological processes, or groups of interrelated processes, known to be active in and on the Earth's crust and thought to be relevant to the assessment of the long-term stability of a radwaste repository. The theme of Part III is the application of this knowledge in the endeavour to determine the safest possible disposal system for the waste type under consideration, i.e., the one most likely to guarantee that any consequent dose to man will not exceed accepted regulatory limits. We focus on the two main areas in which earth scientists are directly involved in the decision-making process: in predicting long-term repository performance within the framework of a safety analysis and in selecting the most favourable repository sites. Neither of these areas can be lifted out of their social context without entering a completely abstract and unreal world. In many countries, convincing proof of safety of radwaste disposal is an integral part of the political process which will determine the future of nuclear power in general, and convincing site selection procedures are a basic necessity for overcoming the inevitable local opposition to repository construction. Hence, both areas are subject to nonscientific pressures and boundary conditions and are to be seen within a broader field of political activity we will call social evaluation. Problems of prediction in the geosciences will be treated in Chapter 14, whilst site selection will be illustrated on the basis of specific case histories in Chapter 15. This twofold subdivision underlines a basic dilemma which exists, and will continue to exist, in the relation between earth science and public policy. On the one hand, a convincing predictive model for performance assessment requires detailed site-specific input data. On the other hand, suitably specific data can only be collected after a site has already been selected. That radwaste disposal has become such a politically controversial issue arises, at least partly, out of this inevitable contradiction in terms.

CHAPTER 14

Predictive Geoscience

I. INTRODUCTION

The long-term safety of a radwaste disposal system can only be demonstrated by predictive modelling of the release, migration, and behaviour of radionuclides in the natural environment. A complete safety analysis consists of two main parts: (1) a geological system or geosphere model (including an oceanographic model in the case of ocean disposal) which simulates the waste isolation system and predicts where, when, and at what rates specific radionuclides will enter the biosphere and (2) a radiological or biosphere model which simulates the migration of radionuclides through the biosphere and predicts the radiation doses to be expected in individuals and populations. Safety is demonstrated by showing that these doses lie below generally accepted limits. In the following, we are mainly concerned with the geological system model, which is the main tool of performance assessment. The large number of physical and chemical factors, the complexity of their interrelationships, and the long time periods involved require a systems approach and the use of computerized mathematical treatments. For our purposes, the geological system model can be thought of as consisting of three main components: (1) the input data, describing quantitatively the properties of the different elements of the system (including waste inventory, waste package and repository design and site characteristics); (2) a system analysis, describing how the different elements and subsystems interact (including mathematical formulations of physical and chemical processes, geological situations and release scenarios); and (3) an estimation of the probability and effects of disruptive events which could negatively affect the system in a nondeterministic way (including seismic and volcanic activity, glaciation and human intrusion).

From the point of view of geology, three types of prediction are associated with the three main components (Fig. 86), and these will be treated in more detail in the succeeding sections. First, the input data must include a description of the geological environment of the repository, including such parameters as the groundwater flow pattern, fracture distribution, permeability, rock

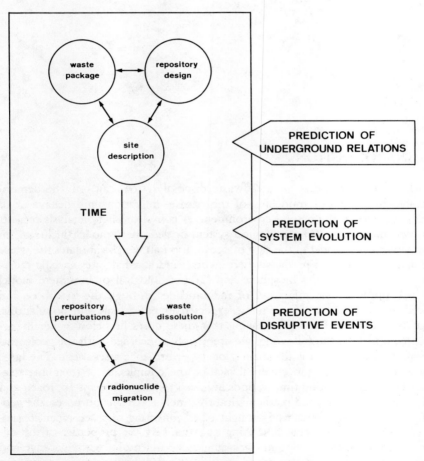

Fig. 86. Schematic diagram of the different types of predictive earth science and their relation to performance assessment.

body geometry and composition and temperature distribution. Many of these parameters have to be deduced from the results of remote sensing, geophysical and geochemical surveying, borehole logging and surface mapping. This implies a considerable degree of interpretation and uncertainty of the same type as encountered in the exploration for subsurface water, energy and mineral resources, and involves skill in predicting underground relationships. Second, the system analysis must be based on a detailed understanding of natural processes and their perturbation by the repository and its contents, usually derived from laboratory experimentation, *in situ* testing, and interpretation of natural analogues (see Part II of this book). This is translated into computer codes governing separable parts of the whole system (subsystems), which are then connected in sequences or feedback networks. Many of these codes have been developed from ones used in civil engineering and environmental studies, with the main aim of predicting system evolution either on a hypothetical basis (generic modelling, essentially mathematical experimentation to explore the importance of the different parameters and to suggest system improvements) or on the basis of a specific situation (site-specific modelling). However, the normal course of system evolution may be disturbed by events which are not determined within the system, but impressed on it from outside, particularly when considering very long periods of time. Thus, the third component of performance assessment is the working out of the probability and effect of changes in the basic parameters and main scenarios, which for the geologist means predicting disruptive events over time spans up to 1 m.y. Although all these aspects of predictive geology are intimately connected, the following discussion is structured under these three headings for the sake of clarity.

II. PREDICTING UNDERGROUND RELATIONSHIPS

How accurately can we predict the properties and distribution of rock masses and fluids underground? This question belongs to what is generally referred to as subsurface mapping, the basis of many types of water, energy, and mineral resource exploration. For radwaste disposal, however, safety is the keyword of the search, rather than profit: geological situations are required which are likely, from general considerations, to provide a safe environment for a repository, and then detailed information is required to build and feed into the predictive model, which is the basis of any safety analysis. Nevertheless, the instrumentation and methods now undergoing development have mainly evolved out of techniques used in the search for water, oil, natural gas, and minerals, joined more recently by those used for studying the migration of pollutants in the subsurface environment and for surveying the ocean floor. A list of techniques relevant to subsurface mapping would include

1. *Remote sensing techniques:* aerial photography; thermal infrared imagery; side-looking radar; and multispectral scanning from aircraft and satellites
2. *Geophysical exploration:* seismic/acoustic methods (seismic reflection profiling, high-frequency echo-character mapping, microseismic/acoustic emission monitoring); mapping of magnetic, gravity, and electrical conductivity anomalies and radar techniques
3. *Borehole logging:* electrical wireline logs such as natural γ, γ density, neutron, pulsed neutron, acoustic, normal resistivity, focused current, induction, dipmeter, electromagnetic propagation, thermal, in-hole television and acoustic televiewer
4. *Hydrological testing and sampling:* monitoring of springs and thermal/mineral-water issues; groundwater monitoring; chemical analysis and isotopic analysis; borehole testing (packer tests, in-hole tracer tests, spinner flowmeter, tube wave surveying, etc.); and permeability determinations
5. *Geological mapping and geostatistics:* profile and section construction; fracture mapping and structural analysis; form surface analysis; and kriging

These groups of techniques are all interrelated in practice, although it is natural to carry out remote sensing and surface studies in advance of physical probing, because drilling and borehole logging is an expensive and time-consuming activity which must be carefully located. As an example of the numerous types of subsurface mapping problems, all of which require a carefully coordinated sequence and combination of techniques, we take the problem of identifying unfractured rock volumes at depth in crystalline rocks, the main aim of disposal site exploration in several countries (Fig. 87).

Fractured zones in hard and impermeable crystalline rocks serve as pathways for rapid fluid transport and as zones of weakness for crustal movements. Although fault zones and volumes of less intensely and more intensely fractured rock are typically unsystematic in distribution, even in very homogeneous rock bodies (see Figs. 62 and 67), most crystalline complexes show a block-like structure, with angular masses of unjointed or poorly jointed rock surrounded by more or less planar zones of jointed or brecciated material, forming a complicated three-dimensional network. Since a mined HLW repository should lie completely within one or more of the unjointed rock masses at depths of 500–1000 m below the surface, and because the number of drillholes which penetrate the unjointed masses must be kept to a minimum (each drillhole is expensive, and also a potential direct connection to the biosphere), the first problem is to delineate such locations with sensing and extrapolation techniques. Although it is possible to adjust repository de-

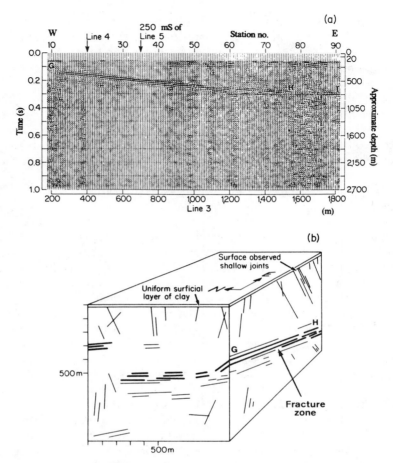

Fig. 87. One of the main methods of subsurface exploration is reflection seismology, using the fact that some types of geological discontinuity reflect shock waves from the surface with travel times related to the depth of the structure. This method is now being adapted to the prediction of fracture zones at depth in crystalline rocks. (a) Seismic reflection profile along an EW traverse through the Lac du Bonnet granite in Manitoba, Canada, the proposed site of Canada's Underground Research Laboratory for radwaste investigations. (b) Block diagram constructed by the interpretation of several such profiles, showing the prominent reflector GH in three dimensions. It is thought to represent a subhorizontal zone of fracturing at a depth of a few hundred metres. (From Mair and Green, 1981, by permission of Macmillan Journals Ltd.)

sign to the actual fracture zone system found during construction, precon-struction data must predict that a sufficiently large volume of suitable rock is in fact present at the required depth. Two types of technique are presently under development to tackle this problem: extrapolation of surface data and reflection seismic profiling (Fig. 87). The first is an extension of structural analysis techniques long in use (statistical analysis of fracture orientation, length, spacing, and type, and analysis of present and past stress fields) and permits the position of prominent, steeply dipping fracture zones to be reconstructed at depth. The second technique, reflection seismology, is being used to identify gently dipping fracture zones which do not have a surface expression but which are commonly encountered in boreholes. At present, both techniques are being developed and tested against actual borehole data (e.g., acoustic televiewer logs) or other subsurface observations such as mines, shafts, underground laboratories and borehole-to-borehole tests, but the aim is to develop a predictive capability in the absence of such data.

The problem of fracture zone identification at depth also points to a more general aspect of predicting underground relationships. The above techniques may allow a preliminary selection of favourable sites and a description of the fracture geometry in their surroundings, but a subsequent performance assessment also requires data on the fluid transport properties of the fractured zones, both in the near field and the far field (groundwater flow rates and pathways), and on the retarding properties of the rock matrix (e.g., sorption on rock and joint minerals and diffusion and dispersion properties). Many of the *in situ* experiments being carried out in underground laboratories are directed towards obtaining typical values for many of these parameters, but whether it is going to be justifiable to apply such values to other sites without control determinations is an open question. Control determinations, however, entail extensive exploratory drilling, which is expensive and introduces new elements of risk. In essence, the problem of predicting underground relationships requires answers to two questions: what is the minimum amount of borehole data required to provide sufficient confidence in the performance assessment, and what is the minimum amount of testing required to provide a sufficiently good estimate of the uncertainties in the subsurface site characterization.

III. PREDICTING SYSTEM EVOLUTION

The second type of predictive geoscience involves the construction of mathematical models of repository systems using simplified analogues of the complex rock, fluid and waste interactions expected from physicochemical theory, laboratory experimentation and field investigations. Much of this book has been concerned with understanding these processes and with illustrating the complexity of the problem at hand. However, certain processes stand out as being of critical importance and system modelling generally proceeds by (a) identifying critical subsystems, (b) creating a computer code to simulate the subsystem in isolation, and (c) combining the various subsystems into logical sequences or interactive networks. For instance, the Dynamic Network model (DNET) developed at the U.S. Sandia National Laboratories to investigate radionuclide release and migration from a HLW repository in a bedded salt formation contains submodels mathematically representing fluid flow, heat transport, salt dissolution, salt creep, thermal expansion, subsidence, and fracture formation and closure. The Swedish waste disposal effort has developed a similar model for granitic environments (Fig. 88) and used it to estimate doses to humans, using several different release scenarios and data from specific sites. A vast amount of human effort and computer time has already been invested in the development, testing, verification, and validation of these and the many other models which are being constructed for repository performance assessment the world over. Using conceptual designs and typical input parameter values, innumerable sensitivity and uncertainty analyses have been carried out and there is general agreement that predictive modelling of waste characteristics, repository perturbations, near-field processes and geosphere migration has already reached a high level of sophistication, often far outstripping the quality of the input data.

As a typical example of this type of work we take NUTRAN, a package of computer programmes developed by a group of U.S. scientists to calculate doses to humans from radioactive materials escaping from deep-mined repositories in groundwater (Fig. 89). It consists of four computer programmes: ORIGEN to calculate the inventory of radionuclides in the waste as a function of time; WASTE to model the release of radionuclides from the repository and their transport through the subsurface to biosphere; BIODOSE to compute the doses to humans of radionuclides released to the biosphere; and PLOT to combine the results of WASTE and BIODOSE. The geological system model, WASTE, consists of a series of subsystems modelling the sequence of events most likely to lead to radionuclide release in a nonsalt environment (release scenario) and incorporating the effects of advective transport, hydrodynamic dispersion and chemical interactions between groundwater, waste components and rock. The calculation begins by calculating the length of time required for the repository to refill with water and to repressurize after closure. This is determined from the hydraulic parameters of adjacent aquifers, using a stream tube model. The three-dimensional flow field is modelled as a network of one-dimensional flow paths (stream tubes), rather than as a lattice of cells (finite element or finite difference method).

Using the results of this first step, or an arbitrary resaturation time, the next step is to calculate the rate at which a specific radionuclide is released to the groundwater and leaves the repository (near-field processes). This is given by the slowest of three processes: leaching of the waste form, turnover of water in the repository room, and solubility of the element in the water

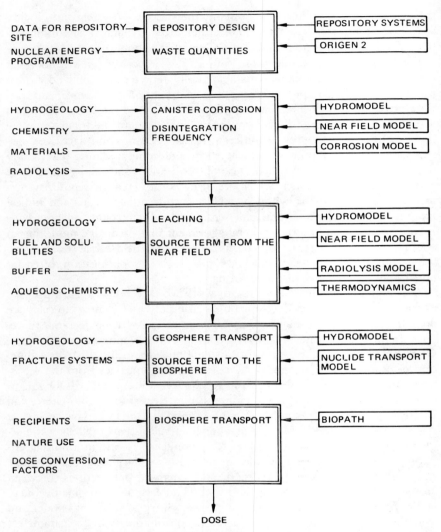

Fig. 88. Flow sheet showing the sequence of subroutines used by the Swedish KBS project for the safety analysis of subsurface HLW repositories in crystalline rocks. The central scenario assumes that the waste canisters are penetrated successively by groundwater during an initial period of 10^5–10^6 years, in a repository at 500 m depth and at least 100 m from the nearest fracture zone. (From Swedish Nuclear Fuel Supply Co., 1983.)

flowing past the waste canister. The result of this calculation, or an arbitrary source term, forms the starting point for the next computation, which models the transport of a dissolved radionuclide in the groundwater. This calculation uses the same stream tube network as before, splitting each flow path into

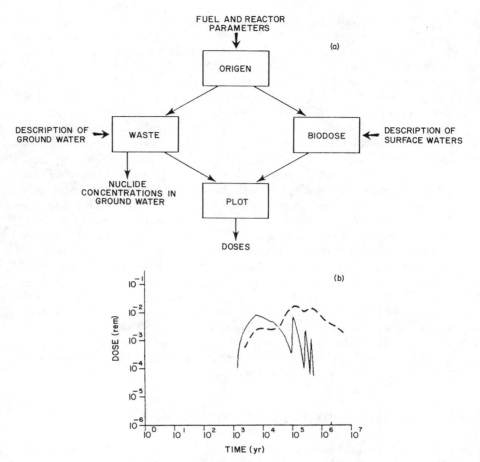

Fig. 89. Example of a model for estimating the long-term hazard connected with the transport of radionuclides out of deep HLW repositories by groundwater. (a) The NUTRAN system of computer programs. (b) Comparison of NUTRAN results (——) with those obtained using a KBS system (– – –) (similar to the one illustrated in Fig. 88), using the same set of input data. (From Ross *et al.*, 1981, by permission of Harwood Academic Publishers.)

segments so that the output of each path segment becomes the input to the next. This approach also allows the modelling of small flaws, such as unsealed shafts and boreholes or narrow fault zones, avoiding the difficulties such features cause using the finite element and finite difference methods. Sorption is modelled by assigning retardation factors to the different elements (see Chapter 11, Section IV,A). The final step in the sequence is the calculation of the concentration of the radionuclide in a surface water body or in a well, using a two-dimensional equation for a homogeneous aquifer. Correction factors have to be introduced at various points to take into account the rates of radionuclide decay and daughter nuclide release. The application of the whole computer package to the Swedish HLW concept and a comparison between the NUTRAN results and those obtained using a different set of computer programmes (GETOUT and BIOPATH) is illustrated in Fig. 89.

In general, the work described in this section addresses the question: "What would happen if. . . . " It is not predictive in the sense of forecasting what will actually take place at some future time: it explores the effects of possible or probable geological events and processes on various conceptual repository designs to identify critical features and to improve the quality of waste isolation systems and the various barriers. It identifies weak points and areas needing further research. It moves from the use of typical or theoretical input parameters (generic modelling) towards the application of waste-, repository-, and site-specific data and to the comparison of model predictions with the actual results of field experiments. It investigates the consequences of different release scenarios, sequences of geologically realistic and conceivable events which could result in radionuclide release and migration to the biosphere, in the same way that world models like Global 2000 investigate the effects of various human actions on future resource depletion and environmental degradation. For further examples of this type of modelling, the reader is referred to Figs. 38 and 66. With continuous refinement, such models will eventually be used in the final safety analysis of an actual repository, but their predictive capacity will then be limited by the uncertainties of the input data and by our ability actually to predict the occurrence of future events. We now turn to this latter aspect of predictive geoscience.

IV. PREDICTING DISRUPTIVE EVENTS

Here we are concerned with probabilities (Figs. 90, 91). It is not possible to predict precisely when a particular geological event will take place at some time far in the future. It is often possible to estimate the probability of the event occurring at least once within a given period of time (see Chapter 4, Section IV,B and Chapter 10, Section III,B). A measure of risk is obtained by multiplying the probability of event occurrence by some measure of its negative consequences (e.g., increased quantities of radionuclides released as calculated by the geological system models described above). Evaluation of the probability of geological events of a disruptive nature is a largely subjective procedure based on expert opinion and analysis of the geological record at the present state of knowledge, and as such should also be accompanied by an evaluation of the uncertainties in the estimates. In fact, the assessment of risk requires a probabilistic approach on all levels; the input data will always be in the form of probability distributions because of the uncertainties inherent in site characterization and event probability estimation.

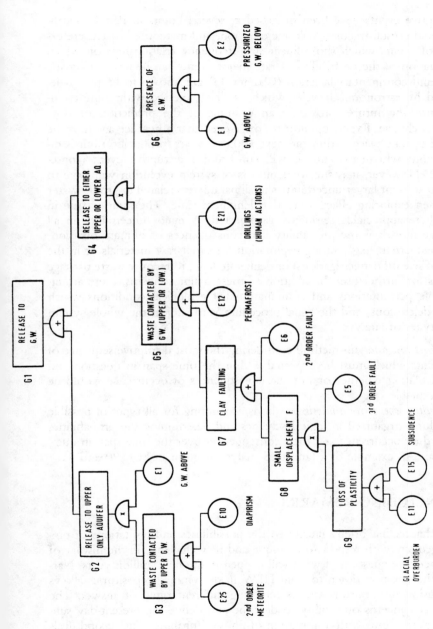

Fig. 90. A fault tree used to calculate the probability of radionuclide release from a deep repository in shale (e.g., Mol site, see Figs. 41 and 49). (From d'Alessandro and Bonne, in de Marsily and Merriam, 1982, by permission of Pergamon Press Ltd.)

Disruptive events have been discussed at several points in Part II. Earthquakes and crustal fracture, volcanic eruption, and meteorite impact, are examples of events which show linear size–frequency relationships on which an estimation of the probability of rare major events can be based. Climatic change and continental glaciation (Chapter 13) are known to be partly determined by astronomical cycles which can be projected with some confidence into the future, providing an example of the prediction of slow geological change. Expert opinion is now agreed that a new ice age must be assumed to take place within the next 100,000 years (probability) but is divided about whether or not it will constitute a disruptive event (consequences). However, it is sure that models of system evolution will have to take into account larger uncertainties in all parameters related to groundwater flow when exploring effects over that length of time. The same is true in areas where appreciable seismicity is expected. A major uncertainty in all predictive models is the probability and consequences of human intrusion, either inadvertent (e.g., during exploration for other raw materials or in the course of scientific investigations) or deliberate (e.g., to recover waste package materials for further use). In addition to these major concerns, there are innumerable permutations and combinations of geological conditions which may be deleterious, and the usual procedure is to subject the whole system to two types of analysis:

1. *Worst case analysis:* method considering the most disruptive sequence of events which must be expected within the time span in question and calculating whether or not the consequences of occurrence would be acceptable
2. *Fault or event tree analysis:* methods accounting for all types of possible failures organized in logical schemes and determining the probabilities of their occurrence and their consequences over the time span in question. An example of a fault tree analysis is given in Figs. 90 and 91.

V. CONCLUDING REMARKS

This chapter has briefly illustrated the possibilities and limitations of predictive geology with respect to a unique and new challenge in the history of the science: the question of what will happen in the next million years. Narrowing the question down to solid HLW in deep-mined repositories allows a subdivision into various aspects requiring different types of answer. The first aspect concerns our ability to describe and define the present-day subsurface relationships with a minimum of physical probing. The second deals with our ability to develop mathematical models to simulate present and future processes in and around the repositories in order to study possible ev-

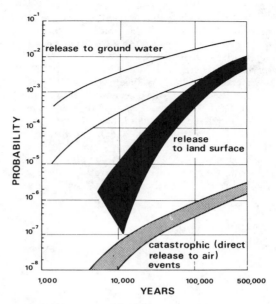

Fig. 91. Top event probability bands for different release scenarios from a shale repository, based on the fault tree analysis of each scenario (for top event, release to groundwater, see Fig. 90). (From d'Alessandro and Bonne, in de Marsily and Merriam, 1982, by permission of Pergamon Press Ltd.)

olutionary schemes for the complex geological system and to evaluate the consequences of changes in the various parameters. The third addresses the problem of evaluating the likelihood of disruptive changes due either to rapid catastrophic events, slow continuous processes or human action. All three aspects involve uncertainties which also need quantifying: uncertainties in the input data and uncertainties due to incompleteness or errors in the models. Assessment of future risk can therefore only be made using a probabilistic approach.

SELECTED LITERATURE

Predictive geoscience and risk assessment: Giletti *et al.*, 1978; Bignoli and Bertozzi, 1979; Carter *et al.*, 1979 (pp. 919–1212); McCarthy, 1979 (pp. 501–554); Northrup, 1980 (pp. 689–858); Webb, 1980; Berg and Maillie, 1981; Klingsberg and Duguid, 1982; Lutze, 1982 (pp. 619–630); Marsily and Merriam, 1982; Coordinating Group for Geological Disposal, 1983 (pp. 83–93); Kocher *et al.*, 1983.

Prediction of underground relationships: Lawrence Berkeley Laboratory, 1978; Lawrence Livermore Laboratory, 1978; Carter *et al.*, 1979 (pp. 955–982); Doctor, 1979; Organization of Economic Cooperation and Development (OECD), 1979c; Everett, 1980; Hobson,

1980 (pp. 189–272); Goetz and Rowan, 1981; Graebner *et al.*, 1981; Mair and Green, 1981; Majer *et al.*, 1981; National Cooperative for Radioactive Waste Storage (NA-GRA), 1982; Smith, 1982 (p. 412–426); Zoback and Anderson, 1982; Damuth *et al.*, 1983; Ramirez, 1983; Kovari, 1983.

Prediction of system evolution: DeMier *et al.*, 1979; Ross *et al.*, 1979, 1981; Brawner, 1980 (pp. 245–269); Campbell *et al.*, 1980; Washburn *et al.*, 1980; Moore, 1981 (pp. 569–606); OECD, 1981b; Petrie *et al.*, 1981; Verkerk, 1981; Baca *et al.*, 1982; Foley *et al.*, 1982; Lutze, 1982 (pp. 765–774); Topp, 1982 (pp. 473–480); International Atomic Energy Agency (IAEA), 1982b (pp. 783–798); Brookins, 1983a (pp. 473–565); Swedish Nuclear Fuel Supply Co., 1983 (pp. 20:1–20:45); NAGRA, 1983b; Pigford, 1983; Wang *et al.*, 1983.

Prediction of disruptive events: Crowe, 1980; D'Alessandro *et al.*, 1980; Lawrence Livermore Laboratory, 1980b, 1982; Bürgisser and Herrenberger, 1981; Cameron, 1981; Christie, 1981; Mallio and Peck, 1981; Bertozzi and D'Alessandro, 1982; Kocher *et al.*, 1983.

CHAPTER 15

Repository Site Selection

I. INTRODUCTION

Performance assessment is essentially site-specific—it forms part of the safety analysis which should prove that a particular site will at all times in the future satisfy the stipulated safety requirements. It does not address the question of how the particular site was chosen in the first place. It is the end of a long line of decisions which is collectively called the site selection process. How these decisions have been and should be reached is the subject of this final chapter. First, we set up a conceptual framework based on the ways in which exploration for all types of underground or marine resources is conducted and, to a lesser extent, on the planning of current radwaste management programmes. Then, we compare this with some case histories to illustrate specific points (Subseabed Disposal Program, Sweden, Switzerland). Since site selection for radwaste repositories is known to be one of the most politically explosive aspects of nuclear energy development, this chapter is mainly concerned with the relation between geology and public policy. Only careful attention to the very early stages of site selection can avoid the situation in which the geologist and his scientific endeavours come under intense political pressures. This derives from the fact that a local population is primarily interested in the reasons why a site "in my backyard" has been selected from what looks like innumerable possibilities. Thus, the emphasis in site selection in populated areas is not, as one would expect, on proving that a particular site is the most favourable, but on proving that other sites must be excluded and need not be investigated further. Selection criteria must be balanced by equally valid and generally accepted exclusion criteria if a rational site selection process is to be achieved.

II. CONCEPTUAL FRAMEWORK

Site selection for any sort of hazardous waste disposal is a complicated procedure with numerous aspects and interactions: scientific, technological, economic, and political. To facilitate a discussion of the geoscientific aspects, a schematic flow sheet will be used (Fig. 92) which is based on exploration and research planning in related fields. We distinguish four phases of investigation, alternating with steps involving political, economic and technical decisions, collectively referred to as social evaluation. In reality, these phases and steps overlap and run parallel to each other and are connected by feedback mechanisms (scientific and social reassessments) which should make the system fully interactive. How these interactions take place depend to a large extent on the political institutions of the country or countries involved, as will become clear from the case histories presented later. Before outlining these, however, we look at the contents of the phases of geoscientific investigation in a more general and abstract way.

A. Phase 1: Reconnaissance

The starting point of the initial phase of investigation is a definition of boundary conditions, i.e., a delineation of demographic, administrative or physiographic boundaries around those areas of the Earth's crust within which a site must be located for other than geological reasons. In addition, a waste management strategy must be established, containing general directives and guidelines (e.g., containment versus dispersal and surface versus subsurface), forecasts of waste type, volume and isolation time, and engineering concepts based on experience in other fields. This initial statement forms the background to phase 1 reconnaissance and screening, which is based on information already available: geological and hydrogeological maps, seismic data and borehole logs from other exploration activities, remote sensing, and geological survey and university publications. The aim of phase 1 is twofold: to identify possible target areas for future detailed and directed study (development of selection criteria) and to reject certain areas as unworthy of more detailed study on geological grounds (development of exclusion criteria). The difficulty in attaining this aim has been, and still is, the lack of a body of past experience, making the criteria used to accept or reject particular areas or geological situations largely value judgements based on experience in other

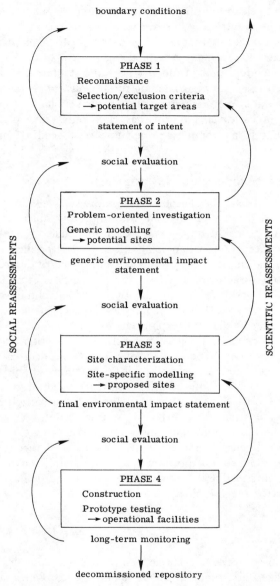

Fig. 92. Hypothetical flow sheet as a framework for discussing the geoscientific aspects of site selection.

fields. This fact is clearly illustrated by the differences in selection and exclusion criteria from country to country. However, the recent history in a number of countries indicates that the development and application of clearly defined and generally accepted criteria at this stage is of utmost importance. Ideally, phase 1 should be concluded with some statement of intent delineating the chosen target areas, explaining the reasons for the choice, and outlining future research requirements, which is then subjected to scientific and social evaluation. Unfortunately, this has often only taken place in cursory discussion leading to an unpublished decision to proceed to phase 2.

B. Phase 2: Problem-Oriented Investigation

The target areas delineated in phase 1 are now subjected to detailed study, planned and carried out with the radwaste disposal problem in mind. Concomitant with this runs the development of more detailed waste management strategies and concepts (see Chapter 2) and of generic predictive models (see Chapter 14) whose aim is to explore the possible effects of different processes, using generalized parameters based on the geological system being investigated. Phase 2 studies generally use already available techniques, such as detailed field mapping, airborne geophysics, reflection seismic profiling, groundwater tracer studies and exploratory drilling, aimed at characterizing the target areas in depth and at developing and refining criteria for identifying potential sites. Because of the breadth of geoscientific data collected in this phase, selection of potential sites and exclusion of unfavourable sites tends to be subjected to closer scientific scrutiny and tends to be based on some sort of scientific consensus. Since the responsibility for site selection rests squarely on the geologist's shoulders at this stage, scrutiny, controversy, peer review and consensus within the profession become important parts of the decision-making process. Typically, phase 2 ends with detailed reports and recommendations, possibly within the framework of a generic environmental impact statement, intended as a major contribution to the public debate and to an institutionalized social evaluation process.

C. Phase 3: Site Characterization

The step from a number of well-defined potential sites to the choice of an actual repository location may be a mainly political decision, but it is accompanied by scientific work aimed at the precise characterization of specific sites. There is now a direct relation between detailed field investigations (e.g., borehole logging and *in situ* testing), laboratory experimentation (e.g., petrographic description, groundwater chemistry and leach and sorption behaviour) and sophisticated mathematical modelling using site-specific data. The aim is to determine whether a potential site does in fact show its supposed ability to prevent radionuclide release to the environment over the time period and within the dose limits stipulated. It can also be coupled to an actual selection of the scientifically most favourable site as a contribution to the final decision-making process. Phase 3 also results in the refinement of established techniques or the development of new techniques with special reference to the problem of radionuclide immobilization, dissolution and migration. The data from these investigations are incorporated in site characterization reports, which may be an important part of the final environmental impact statement.

D. Phase 4: Construction

Social pressures concentrate around phase 3 because of the "not in my backyard" syndrome and the basic political dilemma of majority rule versus minority rights. The social evaluation at the end of phase 3 faces stark political and financial realities which may completely override the scientific merits of the case. If the ground has been carefully prepared in phases 1 and 2, however, a solution will be found to allow work to proceed to phase 4, the period of construction, first of prototypes or demonstration facilities, later of the actual repository, accompanied by monitoring programmes which document the actual perturbations caused by the excavation of the repository and, later, by the emplacement of the waste.

This flow sheet (Fig. 92) is not in any way intended as an ideal plan, but rather as an imperfect representation of a complex interacting system (industrial democracy) which recognizes the need for a logical scientific approach to site selection and, at the same time, the requirement that this must take place within a rational, nonscientific framework. Its aim is merely to provide a convenient shorthand for discussing particular case histories. For example, the selection of many U.S. shallow land burial sites for LLW (see Chapter 5, Section IV) bypassed several steps between phase 1 and phase 4 during the 1950s and 1960s. During the 1970s, however, there was a whole series of reassessments, both scientific and social, which led, first, to a redefinition of the initial boundary conditions (climatic zones, state compacts, etc.) and, second, to the setting-up of more conscious procedures for the development of the next generation of repositories (regulatory procedures and environmental impact statements). Scientific investigations are now again in phase 2 (with an enormous input of experience from the aborted phase 4 facilities), and social evaluation procedures are under review, stimulated by the recognition that hazardous wastes of all types are increasing enormously in quantity and that present adversarial decision-making exacerbates the inevitable conflicts rather than channels public involvement towards the attainment of mutually acceptable solutions. In contrast to LLW site selection, the search for subsurface HLW sites worldwide has hardly progressed past phase 2. As illustrative examples, we take the search for subsurface HLW repository sites in crystalline rocks in Sweden and Switzerland. First, however, we turn to a programme which at the moment is less constrained by the political dimension and therefore more closely follows a step-by-step procedure like the one outlined above: the Subseabed Disposal Program.

III. CASE HISTORY: SUBSEABED DISPOSAL PROGRAM

The Subseabed Disposal Project began in the late 1960s as an oceanographer's pipe dream and has grown since 1973 into an international, multidisciplinary programme which presents a serious alternative to the land-based national HLW disposal concepts. What began as a simple, catchy idea—that vast areas of the deep ocean floor are blanketed with sediments which should make ideal enclosing media and which lie far from centres of population in the most stable and predictable environment known on Earth—has now developed into a wide-ranging research and development effort, financed collectively by 11 countries (see Chapter 11, Section IV,F). Scientists from these countries comprise the Seabed Working Group, formed in 1977 under the auspices of the OECD Nuclear Energy Agency. Unlike many land programmes, the development of subseabed disposal research has been based from the beginning on peer review at many different levels, so that it is governed by the normal assessment and consensus-building processes that are the hallmark of science. This reflects a position free from the pressing political demands which are so characteristic of the land-based disposal programmes. Phases 1–3 of the site selection process have been, and will be, allowed to proceed with continuous scientific evaluation and logical screening, up to the point at which a particular site or several sites can be pinpointed precisely and shown with appropriate predictive models to meet recognized safety requirements. At this point, the programme will enter an extremely difficult and controversial phase of social evaluation, concerning the international legal and control aspects of the exploitation of the ocean floor. At the moment, however, funding is assured, at least to the end of phase 2, and the programme is proceeding on the assumption that the legal problems will eventually be solved. In the following section, we review past and present research planning with respect to site qualification.

The Subseabed Disposal Program started out with clearly formulated criteria for delineating favourable areas of the deep ocean floor (self-imposed boundary conditions for phase 1). The preliminary assessment limited the ocean areas under consideration to those thought to show a high degree of stability (long and continuous sedimentation, minimal seismic or volcanic activity, and low probability of oceanographic or tectonic change) and a thick sequence of sediments with favourable barrier characteristics (low permeability, high sorption coefficients, constant properties over wide areas, and favourable mechanical properties). Of lesser importance were social considerations such as the absence of valuable natural resources (commercial fishing, manganese nodules, etc.) and remoteness from other human activities (such as surface shipping lanes). Application of these preliminary criteria indicated two main environments, the abyssal hills and the distal abyssal plains, in mid-plate/mid-gyre situations, as the ones most likely to provide suitable conditions. Phase 1 (1973–1976) consisted of a more detailed assessment of the idea on the basis of published and archived data and of a first estimation of technical and environmental feasibility. At the end of this phase, the period of scientific evaluation resulted in the selection of five target regions in the northern Pacific and Atlantic oceans (Fig. 93). At the same time, detailed planning for phase 2 and the organization of the programme on an international basis took place. The results of phase 1 were so convincing, and the need for an ocean alternative so pressing, that by 1981 the Seabed Working Group was fully operative, with a total funding of the national subprojects amounting to $15 million.

Given continued financial support (at the moment, there is no industry support, and government grants result from competition with other research projects), phase 2 is planned to be completed by 1986–1988. For each target region, an iterative sequence of field studies has been initiated, becoming more specialized as the results from generic study areas (10^4–10^6 km^2) allow the definition of specific study areas ($< 10^4$ km^2), possibly resulting in the identification of potential candidate sites (Fig. 93). For each generic study area, four phases of oceanographic surveying are envisaged and have been carried out in some cases: (1) reconnaissance marine geology and geophysics (bathymetry, seismic profiling, magnetics and heat flow); (2) detailed topographic mapping using swath-mapping techniques (side-scanning radar and Sea Beam) to evaluate bottom irregularities such as seamounts and fault scarps; (3) high-resolution near-bottom seismic surveying to delineate the structure of the upper 50–100 m of sediment; and (4) sediment sampling to at least twice the proposed emplacement depth (deep-sea drilling, long piston coring). Some of the results from one of these study areas are given in Figs. 76 and 77. Ideally, after each phase, the data will be evaluated and the course of subsequent research reviewed in the light of it, although some parallel studies and reversals of sequence may be unavoidable for organizational reasons. Also, the results of other work by SWG task groups, including biological and physical oceanography, systems analysis and engineering field studies, will be continuously incorporated in the site qualification process.

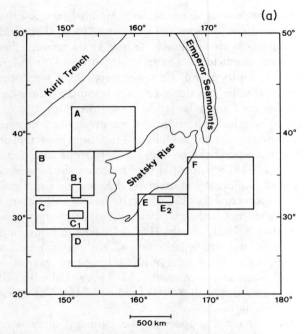

Fig. 93. Site qualification within the framework of the Subseabed Disposal Program. (a) Location of study areas within study region PAC I (northwest Pacific Basin, see Figs. 75–77) selected for continued assessment. (b) Study location E_2 within study area E. E_2 was subjected to detailed investigations during the 1980 cruise of the oceanographic research vessel *Vema* (piston and gravity coring, heat-flow measurement, acoustic character mapping, etc.) and was a deep-sea drill site in 1982 (DSDP 576, see Fig. 77). (From Shephard, 1983.)

The aim of phase 2 is to characterize representative study areas and to identify potential candidate sites. The following phase 3 (end 1980s to end 1990s) will be directed towards demonstrating engineering feasibility (emplacement, transport and recovery systems) and towards achieving international legal acceptability (amendment of the London Convention, establishment of an international control body and agreement on safety standards). If a decision on legal acceptability can be reached, it is estimated that the demonstration of a capability for safe HLW disposal will require a further decade, so that routine use of the technique could start by 2010 A.D. at the earliest.

IV. CASE HISTORY: SWEDEN

Site selection for underground HLW repositories in Sweden is at present in phase 2 and current investigations are expected to be completed by the late 1980s. In contrast to the Subseabed Disposal Program, however, it is

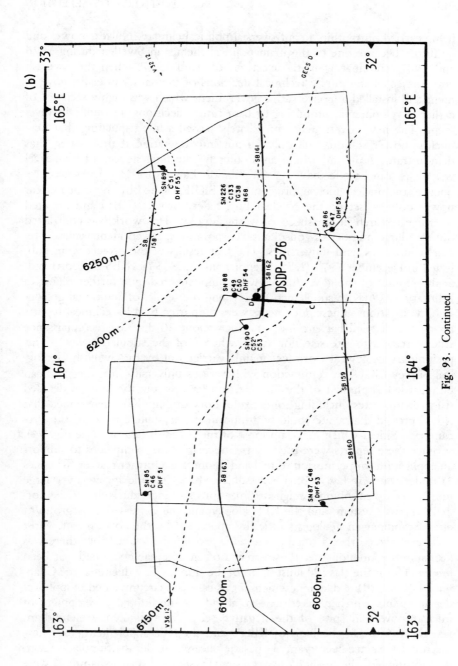

Fig. 93. Continued.

being carried out within a controversial political climate. Nuclear power and its future became one of the main political themes in Sweden during 1976 and 1977, culminating in the elections of April 1977, when the Swedish voters rejected the purely technical decisions of the earlier Socialist government and installed a Conservative government which was highly sceptical of earlier developments, promising to base future decisions on public involvement. The new government immediately passed a law stipulating that new nuclear power stations will only be put into operation if the owner "has demonstrated how and where an absolutely safe final storage of high level waste . . . can be effected . . ." (Stipulation Law, April 1977). As a consequence, the nuclear fuel supply company, SKBF, owned by the four nuclear power utilities, set up the Nuclear Fuel Safety project (KBS) and initiated intensive research in all aspects of waste disposal. This work was contracted out to a large number of consultants, corporations, government facilities, research institutes, and universities. The first results of the project were published in December 1977 (KBS-1 report) and June 1978 (KBS-2 report) and, at the request of the government, were supplemented by a further report in February 1979 (KBS-supplement) describing the results of additional geological investigations. These documents were supported by 120 technical reports giving the detailed background, of which about 20 describe reconnaissance studies related to site selection (the "where" of the stipulation law). This preliminary geological work was mainly carried out by the Swedish Geological Survey (SGU) in conjunction with various university institutes and can be regarded as phase 1 of the site selection process. On the basis of detailed study at four sites, including some exploratory drilling, KBS maintained that it had proved that a site could be found in the crystalline rocks of the Precambrian Shield which make up most of the Swedish bedrock (see Fig. 94).

Since these reports were used by the nuclear power companies to support an application for permission to fuel and to operate a further reactor (Ringhals 3) under the new law, KBS-1 was subjected to scientific review by a large number of Swedish and foreign organizations and individuals under the normal "remiss" system, and the KBS-supplement was reviewed by a group of eight Scandinavian geologists officially appointed by the government. Most reviewers agreed that a suitable site *could probably* be located, but there was a unanimous opinion that the existence of an acceptable site had not been *proven*. This is the classical illustration of the siting–safety dilemma (see Chapter 15, Section I), a dilemma which in this case was circumvented by political developments which resulted in a go-ahead for the limited development of nuclear power in spite of the negative scientific assessment (referendum, March, 1980). On the basis of KBS-1 and the KBS-supplement, four new reactors have since been granted operating licenses under the Stipulation Act. Applications for the fuelling of two more (Forsmark 3, Oskarshamm 3) were

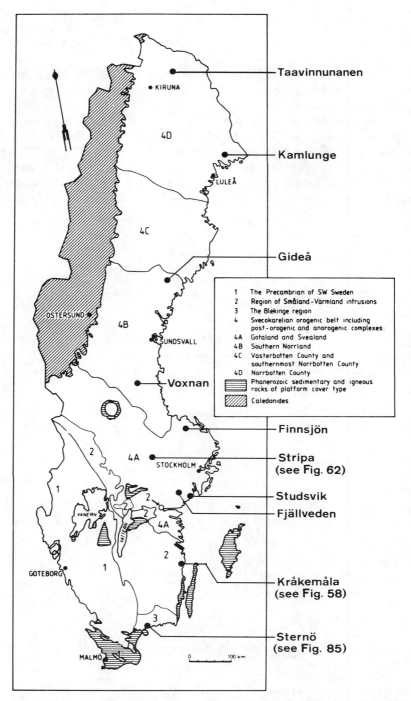

Fig. 94. Outline geological map of Sweden, showing the location of areas where research on the geological aspects of radwaste disposal is being or has been carried out (From Fromm *et al.*, 1980.)

submitted in 1983, supported by a new, more detailed report on the "how" and "where" of HLW disposal (KBS-3, May 1983). The summaries in this report, and the technical reports on which these are based, give a good overview of the structure of phase 2 in the Swedish site selection programme.

The investigations envisaged during phase 2 of the Swedish programme are summarized in Fig. 95 and are expected to be completed by 1990. The data collected during phase 1 (i.e., up to 1980) and the critical evaluation of that work by the geological reviewers have been fed back into the research programme in several ways, including a tightening and upgrading of the study

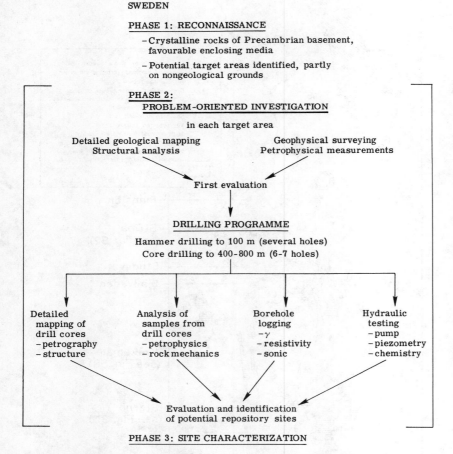

Fig. 95. Summary of the standard programme for phase 2 site investigations of the Swedish KBS project. (From Thoregren, 1983.)

site field investigations and a clearer subdivision of responsibilities. In addition to further investigations at the four original sites, several new sites are being studied, some by a private company, Swedish Geological, under contract from the SKBF, some by the Geological Survey under contract from the Swedish government. This setup resulted from the social evaluation after phase 1, which showed that a clearer separation between the radwaste "producers" and the governmental control and regulating bodies was highly desirable. Investigations at the new sites now include extensive hammer drilling (10–30 drill holes up to 250 m deep) and between 5 and 15 carefully located, oblique, core boreholes down to depths of 600 m. By 1990, sufficient data should have been collected and evaluated that two or three sites can be selected for further detailed investigation, including tests in underground facilities like the present facility in the Stripa mine. However, the basic problem of exclusion criteria will remain, i.e., with such huge areas of superficially similar crystalline rocks, it will be difficult to give a convincing answer to the question "Why not somewhere else?" In a country with a local land use veto built into its legal structure, this is likely to be the critical question.

V. CASE HISTORY: SWITZERLAND

As a final example of site selection problems, we turn to Switzerland, which has been struggling to establish appropriate procedures over the past 10 years. Since 1972, radwaste disposal research and development has been directed by a national cooperative company, NAGRA, financed by the "waste producers," i.e., the private companies operating or planning to operate nuclear power stations and the confederate government (to the extent that it is responsible for LLW originating in hospitals, research facilities, etc.). The NAGRA acts as a central coordination and planning institution, contracting out the basic research to private companies, universities and government-operated research facilities. Site selection activities are complicated by two sets of circumstances which make progress much slower and more laborious than the same activities in Sweden, with the result that investigations have hardly progressed past phase 1. First, Switzerland is a confederation with complicated legal and political rulings governing the sharing of confederate, cantonal and communal authority. Connected with this is the fact that the country does not have a Geological Survey (exploration and exploitation of mineral resources lies in the jurisdiction of the cantons) and that geoscience research is distributed unsystematically between university institutes, private geotechnical firms, scientific commissions and cantonal offices. Hence, there was no obvious organization for NAGRA to consult in planning and carrying out the phase 1 reconnaissance. Second, the physical boundary conditions are generally unfavourable for HLW disposal, judged on a worldwide basis. The country lies near the European–African plate boundary, which was an active collision zone up to 20 m.y. ago (formation of the Alps) (Fig 96) and shows evidence of appreciable crustal movement up to 5 m.y. ago (formation of the Jura Mountains). The present climate is humid temperate (supporting fertile soils in lowland regions, and hence a dense population), but abundant evidence indicates that the country was covered by thick glaciers derived from the Alps about 12,000 years ago, a situation which is expected to repeat itself in the, geologically speaking, near future (Chapter 13). These and other factors such as seismicity and deep groundwater flow meant that NAGRA's job from the beginning was not to find the most favourable site, but to find the least unfavourable one, and then to prove that, even so, it would be safe.

(a)

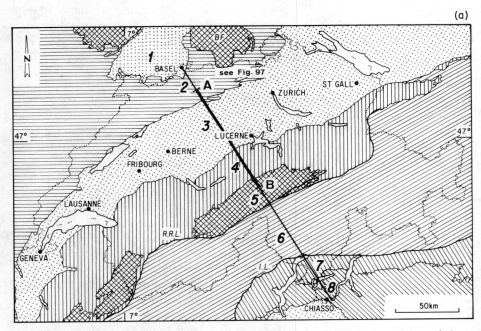

Fig. 96. Geological framework of the main target area for HLW disposal research in Switzerland. (a) Outline map showing the position of the target area to the south of the Black Forest (see Fig. 97). Line shows the position of the Swiss Geotraverse; segment A–B is shown in Fig. 96b. Key: 1. Rhine graben fill; 2. Jura Mountains; 3. Molasse Basin; 4. Helvetic Nappes; 5. External Massifs; 6. Pennine Zone (collision suture) and Austroalpine Nappes; 7. South Alpine basement; 8. South Alpine cover. Other symbols: BF, Black Forest; R.R.L., Molasse Basin between the Jura Mountains and the Aar external massif. (From Trümpy *et al.*, 1980, by permission of Wepf Verlag.)

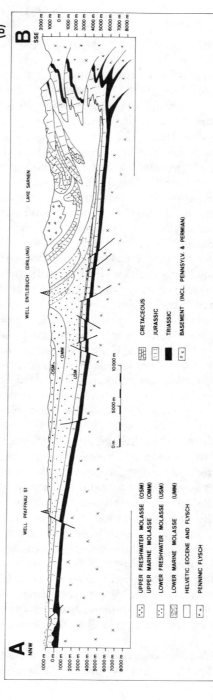

Fig. 96. Continued.

Because of the uncertainty as to who was responsible for issuing permits, and because of the public outcry when it was discovered that field work and test drilling were being carried out surreptitiously, geological investigations during the early part of phase 1 (1972–1978) were piecemeal and unsystematic. Nevertheless, in a preliminary statement on all aspects of radwaste management in February 1978, options had been narrowed down areally to the northern part of Switzerland (Jura Mountains, northern Swiss plain) and geologically to several possible enclosing media (bedded salt, anhydrite, shales or marls or crystalline rocks) and/or particular hydrogeological conditions (localities with very old groundwater). Political developments at around this time resulted in a clarification of the legal aspects of site investigations (competence was transferred to the confederate authorities, with the power to preempt local decisions, if necessary) but also put a time limit on phase 1 operations. Like the Swedish Stipulation Law, the new ruling set a deadline at 1985, by which time potential repository target areas for all radwaste categories should be defined, and corresponding detailed construction plans and generic safety analyses submitted. If safe disposal of the radioactive wastes could not be "guaranteed" by this time, the operation permits of the nuclear power stations would be rescinded. The pressure of this time limit encouraged internal screening by the NAGRA and its consultant geologists, rather than the attainment of a broad consensus. Only 2 years later, proposals were submitted for the sinking of 12 deep boreholes in an area south of the Black Forest (Fig. 97), which immediately became the centre of controversy. According to the project description which accompanied the detailed permit requests, the crystalline basement underlying the Mesozoic successions which blanket the area was to be the main object of investigation, and the area had been chosen because of the shallow depth of the top of the basement (< 1000 m), the lack of faulting, and the very low seismicity.

Perhaps the most striking aspect of this project is the enormous effort which is being put into what is essentially a reconnaissance study, probably under the pressure of the politically determined deadline. If the early part of phase 1 took place behind the scenes, uncontrolled by peer review, the second part (1978–1984) is taking place almost explosively, involving hundreds of earth scientists and support workers. With the completion of each borehole, a flood of data is generated and stored, together with parallel running reflection seismic profiling, hydrogeologic testing and sampling and laboratory investigations. A satisfactory synthesis of this data would represent many years of scientific endeavour. However, a synthesis is required by law by late 1984 and there is some doubt as to the reliability of conclusions reached at such a breakneck pace.

Information at present available, after the completion and preliminary logging of four boreholes which penetrate variable thicknesses of Mesozoic sed-

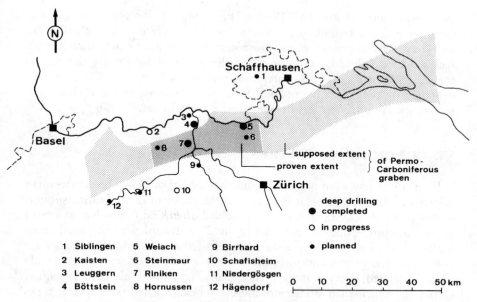

Fig. 97. Location of the deep exploration drill holes in the target area for HLW disposal in northern Switzerland. The drilling activity and associated reflection seismic profiling proved the existence of a deep graben in the supposed crystalline basement containing thick Permo-Carboniferous sediments (including economic coal deposits). (Data from various National Co-operative for Radioactive Waste Storage publications.)

imentary rocks and ~ 1000 m of the underlying basement, is hardly encouraging and points to the need for a scientific reassessment of the whole programme. The present target area has turned out to be much less appropriate than originally supposed. Only one borehole encountered homogeneous granite, of otherwise unknown size and extent, whereas two penetrated thick successions of Permo–Carboniferous sediments (including thick coal seams) below the Triassic unconformity. The extent of the Permo–Carboniferous graben is now known from the reflection–seismic profiling, carried out after, rather than before, the borehole sites were chosen. It traverses the whole target area in a broad band presumably associated with large, hidden, east–west fault zones. In addition, there is evidence for very deep water circulation in the basement, with artesian conditions bringing water to the surface as thermal and mineral springs throughout the area. Even if the results encouraged a continuation of this line of investigation, the blanketing effect of the Mesozoic formations, which, in contrast to Sweden, prevent direct field mapping of lithologies and fracture zones in the basement, will probably mean that phase 1 of the Swiss HLW repository siting programme would be expected to extend into the 1990s. The results of this enormous conflict between political boundary conditions (proof of safe disposal of HLW by 1984) and scientific conclusions (abandonment of project or continuation with reasonable and flexible time limits for the proof of safety) are difficult to foresee. Add to this the poor preliminary planning by the NAGRA and the existence of a large degree of local land-use autonomy, and it is clear how repository site selection can develop into a major national crisis.

VI. CONCLUDING REMARKS

From these few examples, and from the history of radwaste management efforts in other countries, it is clear that the problem of repository siting is highly political. The earth scientists involved often find themselves as pawns in an economic–political game, to be used or discarded as the need arises with little regard for the scientific merits of the case. Most geologists are convinced that careful and methodical study would reveal sites which would satisfy accepted safety standards, but many are uneasy about the political climate in which such studies are presently being carried out. Several types of uncomfortable situation can be distinguished, with many nuances. First, we have the *fait accompli* in which repository siting is carried out with little or no scientific consultation (e.g., U.S. shallow land burial and uranium mill tailing sites) (see Chapter 5, Sections IV and V), and the earth scientist is called in later to assess the extent of the damage and to suggest remedial action. He finds himself concerned with clearing up a mess which might have been avoided if he had been consulted in the first place. Second, we can recognize a "soft money" situation in which large amounts of money are suddenly made available for sophisticated basic research but in a political atmosphere in which there is little guarantee of long-term continuity. The Subseabed Disposal Program, which has the potential of being a systematically and logically implemented project because the target area, the ocean floor, lies in no one's backyard, suffers from this financial uncertainty, which is exacerbated by political manoeuvring in favour of land disposal and by the possibility that it may be blocked by an international legal impasse (see Chapter 15, Section III). Third, we have the situation we can call the "line of least resistance," in which the target area is chosen because it lies in a territory which belongs to the government or to the nuclear power companies, and therefore will arouse a minimum of local opposition (e.g., Hanford Reservation and the Nevada Test Site in the United States) (see Chapter 8, Section III). According to many generally accepted criteria, the area may be geologically quite unsuitable, and the geologist is given the job of finding the least unsuitable site within it and of collecting the data which will be used to prove that it meets the safety standards. Finally, there is the "scientific alibi" situation, whereby the choice of a site or target area is justified on the basis of supposed favourable characteristics, when it is clear to the public that other, nonscientific factors (e.g., access) played a much greater role in the choice. Although most real actions are more complicated than the above-mentioned list would imply, the suspicion of an alibi function of science in site selection can usually only be allayed by showing not only that a site is favourable according to specified criteria, but also that other sites are unfavourable by the same measure. Unfortunately, the alibi suspicion is often strengthened by the "line of least resistance" taken in many countries, since this implies that the geological characteristics of an area are of minor importance in site selection, i.e., that anywhere will do. Whatever the nuances at a particular time and place, these types of situation represent the negative side of what can be a stimulating geoscientific activity. It is fair to say that involvement in the site selection process presents the geological profession with a strong challenge to the maintenance of scientific integrity.

SELECTED LITERATURE

Repository site selection and screening: International Atomic Energy Agency (IAEA), 1977; 1980 (pp. 455–466); Heckmann et al., 1979; Steinborn et al., 1980; Bundesministerium für Forschung Technik, 1981; Organization of Economic Cooperation and Development (OECD), 1981c; de Marsily, 1981; Rogers and Conwell, 1981; Sousselier, 1981; Falconer et al., 1982; Hill et al., 1982; Howell and Olsen, 1982; Morell and Magorian, 1982; Skinner and Walker, 1982; Wynne, 1982; Colgalzier, 1982.

Case history: Subseabed Disposal Program: Anderson, 1979; Barkenbus, 1979; Anderson, 1982; Kelly and Shea, 1982; Shephard et al., 1983 (pp. 272–277); Subseabed Disposal Program, 1983.

Case history: Sweden: Swedish Nuclear Fuel Supply Co., 1977, 1978; Johansson and Steen, 1978; Aberg, 1980; Ahlström et al., 1981; Fromm et al., 1980: de Marsily, 1981; OECD, 1981c (pp. 51–72), 1983b; Paige et al., 1981 (pp. 36–39); de Marsily and Merriam, 1982 (pp. 65–100); Thoregren, 1983.

Case history: Switzerland: NAGRA, 1978; Milnes, 1979; AGNEB/NAGRA, 1980; Fluck et al., 1980; Rybach et al., 1980; Thury, 1980; Trümpy et al., 1980; Bredehoeft and Maini, 1981; Buser and Wildi, 1981; Paige et al., 1981 (pp. 40–44); NAGRA, 1983a.

References

Aberg, B. (1980). "Swedish Investigations on Handling and Final Storage of Radioactive Wastes from Nuclear Power Plants," Lect. transcript, November. ETH, Zürich.

Adam, C. *et al.* (1974). Möglichkeiten des Versenkens flüssiger radioaktiver Abfälle in Gesteinsschichten des tieferen Untergrundes auf dem Territorium der DDR. *Z. Geol. Wiss.* 2, 1017–1031.

AGNEB/NAGRA (1980). Die Endlagerung radioaktiver Abfälle in der Schweiz. *Schweiz. Bauztg.* 98, 1206–1212.

Ahlström, P.-E. *et al.* (1981). Safe handling and storage of high level radioactive waste. A condensed version of the proposals of the Swedish KBS project. *Radioact. Waste Manage.* 1, 57–103.

Aikin, A. M. *et al.* (1977). "The Management of Canada's Nuclear Wastes," Rep. EP 77-6. Ministry of Energy, Mines and Resources, Canada.

Allard, B. *et al.* (1983). "Chemistry of Deep Groundwaters from Granitic Bedrock," Tech. Rep. 83-59. Swed. Nucl. Fuel Supply Co., Div. KBS, Stockholm.

Altenbach, T. J. (1979). "Three-dimensional Thermal Analysis of a High-level Waste Repository," Rep. UCID-17984. Lawrence Livermore Lab., Livermore, California.

American Physical Society (1977). "Nuclear Fuel Cycles and Waste Management." APS, New York.

Anderson, D. R. (1979). "Nuclear Waste Disposal in Subseabed Geologic Formations: The Seabed Disposal Program," Rep. SAND78-2211. Sandia Labs., Albuquerque, New Mexico.

Anderson, D. R., ed. (1982). "Proceedings of the Seventh International NEA/Seabed Working Group Meeting, La Jolla, California, March 15–19, 1982," Rep. SAND82-0460. Sandia Labs., Albuquerque, New Mexico.

Anderson, J. D. (1975). "Radioactive Liquid Wastes Discharged to Ground in the 200 Areas During 1974," Rep. ARH-3093-4Q. Atlantic Richfield Hanford Co., Richland, Washington.

Anderson, R. Y. and Kirkland, D. W. (1980). Dissolution of salt deposits by brine density flow. *Geology* **8**, 66–69.

Andrews, J. T. (1982). On the reconstruction of Pleistocene ice sheets: A review. *Quat. Sci. Rev.* **1**, 1–30.

Argall, G. O., ed. (1979). "Tailing Disposal Today," Vol. 2. Freeman, San Francisco, California.

Augustithis, S. S., ed. (1983). "Leaching and Diffusion in Rocks and their Weathering Products." Theophrastus, Athens.

Baca, R. G. *et al.* (1980). "Coupled Geomechanical/Hydrological Modeling: An Overview of Basalt Waste Isolation Project Studies," Rep. RHO-BWI-SA-82. Rockwell Hanford Operations, Richland, Washington.

Baca, R. G. *et al.* (1982). "Performance Assessment for a Repository in Basalt: A Summary of Preliminary Results," Rep. RHO-BW-SA-248 P. Rockwell Hanford Operations, Richland, Washington.

Bailey, D. K. and MacDonald, R., eds. (1976). "The Evolution of the Crystalline Rocks." Academic Press, London.

Barkenbus, J. N. (1979). "Deep Sea Resources. Politics and Technology." Free Press/Macmillan, New York and London.

Barnes, H. L., ed. (1979). "Geochemistry of Hydrothermal Ore Deposits," 2nd ed. Wiley, New York.

Barraclough, G., ed. (1978). "The Times Atlas of World History." Times Books, London.

Barraclough, J. T. *et al.* (1982). Hydrologic conditions at the Idaho National Engineering Laboratory, Idaho. Emphasis 1974–1978. *Geol. Surv. Water-Supply Pap. (U.S.)* **W2191**, 1–52.

Becker, W. *et al.* (1979). "Properties of Radioactive Wastes and Waste Containers," Rep. NUREG/CR-0619. Brookhaven Natl. Lab., Upton, New York.

Beget, J. E. (1983). Radiocarbon evidence of worldwide early Holocene climatic change. *Geology* **11**, 389–393.

Bell, M. L. and Nur, A. (1978). Strength changes due to reservoir-induced pore pressure and stresses and application to Lake Oroville. *JGR, J. Geophys. Res.* **83**, 4469–4483.

Beneš, P. and Majer, V. (1980). "Trace Chemistry of Aqueous Solutions." Elsevier, Amsterdam.

Berg, G. G. and Maillie, H. D., eds. (1981). "Measurement of Risks." Plenum, New York.

Bertozzi, G. and D'Alessandro, M. (1982). A probabilistic approach to the assessment of the long-term risk linked to the disposal of radioactive waste in geological repositories. *Radioact. Waste Manage.* **3**, 117–136.

Best, M. G. (1982). "Igneous and Metamorphic Petrology." Freeman, San Francisco, California.

Bignoli, G. and Bertozzi, G. (1979). Modelling of artificial radioactivity migration in environment: A survey. *Comm. Eur. Commun. [Rep.]* **EUR-6179**, 1–129.

Bird, G. W. and Fyfe, W. S., eds. (1982). "Geochemistry of Radioactive Waste Disposal." Elsevier, Amsterdam.

Bishop, W. W., ed. (1978). "Geological Background to Fossil Man." Scottish Academic Press, Edinburgh.

Björkman, H. and Trädgårdh, J. (1982). Differential uplift in Blekinge indicating late-glacial neotectonics. *Geol. Foeren. Stockholm Foerh.* **104**, 75–79.

Blasewitz, A. G. *et al.* (1983). "The Treatment and Handling of Radioactive Wastes." Springer, New York.

Blatter, H. *et al.* (1984). "Atlas of Solar Climate." Geogr. Inst., ETH, Zürich.

Bloom, A. L. (1973). "The Surface of the Earth." Prentice-Hall, London.

Bohn, T. and Zoller, P., eds. (1981). "Entsorgung von Kernkraftwerke." Verlag TUV Rheinland, Köln.

Bolt, B. A. (1978). "Earthquakes: A Primer." Freeman, San Francisco, California.

Bolt, B. A. (1982). "Inside the Earth. Evidence from Earthquakes." Freeman, San Francisco, California.

Bolt, B. A. *et al.* (1975). "Geological Hazards." Springer-Verlag, Berlin and New York.

Bornemann, O. (1982). Stratigraphie und Tektonik des Zechsteins im Salzstock Gorleben auf Grund von Bohrergebnissen. *Z. Dtsch. Geol. Ges.* **133,** 119–134.

Bostrom, R. C. and Sherif, M. A. (1970). Disposal of waste material in tectonic sinks. *Nature (London)* **228,** 154–156.

Bott, M. H. P. (1982). "The Interior of the Earth: Its Structure, Constitution and Evolution," 2nd ed. Arnold, London.

Bowen, D. Q. (1978). "Quaternary Geology. A Stratigraphic Framework for Multidisciplinary Work." Pergamon, Oxford.

Bowen, V. T. and Hollister, C. D. (1981). Pre- and post-dumping investigations for inauguration of new low-level radioactive waste dump sites. *Radioact. Waste Manage.* **1,** 235–269.

Bowen, V. T. and Livingston, H. D. (1978). "Radionuclide Distributions in Sediments of Marine Areas used for Dumping Solidified Radioactive Wastes," Rep. COO-3563-62. Woods Hole Oceanogr. Inst., Woods Hole, Massachussets.

Boyle, R. W. (1959). The geochemistry, origin and role of carbon dioxide, water, sulphur, and boron in the Yellowknife gold deposits, Northwest Territories, Canada. *Econ. Geol.* **54,** 1506–1524.

Brace, W. F. (1980). Permeability of crystalline and argillaceous rocks. *Int. J. Rock Mech. Min. Sci.* **17,** 241–251.

Bradley, D. J. (1978). "Leaching of Fully Radioactive High-level Waste Glass," Rep. PNL-2664. Pacific Northwest Labs., Richland, Washington.

Braunstein, J. and O'Brien, G. D., eds. (1968). "Diapirs and Diapirism." Am. Assoc. Pet. Geol., Tulsa, Oklahoma.

Brawner, C. O., ed. (1980). "Uranium Mine Waste Disposal." Am. Inst. Min., Metall. Pet. Eng., New York.

Bredehoft, J. D. and Maini, T. (1981). Strategy for radioactive waste disposal in crystalline rocks. *Science* **213,** 293–296.

Brookins, D. G. (1976). Shale as a repository for radioactive waste: The evidence from Oklo. *Environ. Geol.* **1,** 255–259.

Brookins, D. G., ed. (1983a). "Scientific Basis for Nuclear Waste Management," Vol. VI. North-Holland, New York.

Brookins, D. G. (1983b). Migration and retention of elements at the Oklo natural reactor. *Environ. Geol.* **4,** 201–208.

Brown, P. M. *et al.* (1979). Evaluation of the geologic and hydrologic factors related to the waste-storage potential of Mesozoic aquifers in the southern part of the Atlantic Coastal Plain, South Carolina and Georgia. *Geol. Surv. Prof. Pap. (U.S.),* **1088,** 1–37.

Budd, W. F. and Smith, I. N. (1981). The growth and retreat of ice sheets in response to orbital radiation changes. *IAHS-AISH Publ.* **131,** 369–409.

Bull, C. B. B. (1975). The disposal of radioactive wastes in the Antarctic ice sheet. *Int. Union Geodesy Geophys., Chron.* **102,** 163–167.

Bundesministerium für Forschung und Technik (1981). "Entsorgung. Zwischenergebnisse der Standorterkundung Gorleben." BFT, Bonn.

Burchfiel, B. C. *et al.* (1980). "Continental Tectonics." Natl. Res. Counc.-Natl. Acad. Sci., Washington, D. C.

Bürgisser, H. and Herrnberger, V. (1981). "Zur Risikoanalyse für Endlager in der Nord-schweiz: Ausmass und Wahrscheinlichkeit geologischer Prozesse und Ereignisse," Rep. 430. Eidg, Inst. Reaktorforsch., Würenlingen, Switzerland.

Bürgisser, H. et al. (1979). "Geologische Aspekte der Endlagerung radioaktiver Abfälle in der Schweiz." Schweiz. Energ.-Stift., Zürich.

Burst, J. F. (1969). Diagenesis of Gulf Coast clayey sediments and its possible relation to petroleum migration. Am. Assoc. Petrol. Geol. Bull. 53, 73–93.

Burton, B. W. (1982). "Alternatives to Shallow Land Burial," Rep. LA-UR-82-2532. Los Alamos Natl. Lab., Los Alamos, New Mexico.

Burton, B. W. et al. (1982). "Overview Assessment of Nuclear Waste Management," Rep. LA-9395-MS. Los Alamos Natl. Lab., Los Alamos, New Mexico.

Buser, M. and Wildi, W. (1981). "Wege aus der Entsorgungsfalle." Schweiz. Energ.-Stift., Zürich.

Butlin, R. N. and Hills, D. L. (1982). "Backfilling and Sealing a Repository for High-level Radioactive Waste—A Review," Rep. BRE/CP-6/82. U.K. Dept. of Environment, London.

Byers, F. M. et al. (1976). Volcanic suites and related cauldrons of Timber Mountain-Oasis Valley caldera complex, southern Nevada. Geol. Surv. Prof. Pap. (U.S.) 919, 1–70.

Cahill, J. M. (1982). Hydrology of the low-level radioactive solid-waste burial site and vicinity near Barnwell, South Carolina. Geol. Surv. Open-File Rep. (U.S.) 82–863, 1–101.

Cameron, F. X. (1981). Human intrusion into geologic repositories for high-level radioactive waste: Potential and prevention. Radioact. Waste Manage. 2, 179–187.

Campbell, J. E. et al. (1980). "Risk Methodology for Geologic Disposal of Radioactive Waste: The Network Flow and Transport (NWFT) Model," Rep. SAND79-1920. Sandia Labs., Albuquerque, New Mexico.

Carpenter, D. R. and Martin, R. W. (1980). "Reference Repository Design Concept for Bedded Salt," Rep. UCID-18685. Lawrence Livermore Lab., Livermore, California.

Carpenter, D. W. and Towse, D. (1979). "Seismic Safety in Nuclear-waste Disposal," Rep. UCID-18125. Lawrence Livermore Lab., Livermore, California.

Carpenter, D. W. et al. (1979). "Groundwater Recharge and Discharge Scenarios for a Nuclear Waste Repository in Bedded Salt," Rep. UCID-18119. Lawrence Livermore Lab., Livermore, California.

Carr, J. E. et al. (1980). Geohydrology of the Keechi, Mount Sylvan, Oakwood and Palestine salt domes in the northeast Texas salt-dome basin. Water Resour. Invest. (U.S. Geol. Surv.) 80–2044, 1–60.

Carroll, D. (1970). "Rock Weathering." Plenum, New York.

Carter, M. W. et al., eds. (1979). "Management of Low-Level Radioactive Waste." Pergamon, Oxford.

Carter, N. L. and Hansen, F. D. (1983). Creep of rock-salt. Tectonophysics 92, 275–333.

Carter, N. L. et al. (1977). "Summary Review of Rock Mechanics Workshop on Radioactive Waste Disposal." Memo. Y/OWI/TM-20. State University of New York at Stonybrook.

Carter, N. L. et al. eds. (1981). "Mechanical Behavior of Crustal Rocks." Am. Geophys. Union, Washington, D.C.

Chapin, C. E. and Elston, W. E. (1979). Ash-flow tuffs. Geol. Soc. Am., Spec. Pap. 180, 1–211.

Chavez, P. F. and Dawson, P. R. (1981). "Thermally Induced Motion of Marine Sediments Resulting from Disposal of Radioactive Wastes," Rep. SAND80-1476. Sandia Labs., Albuquerque, New Mexico.

Christiansen, R. L. et al. (1977). Timber Mountains-Oasis Valley caldera complex of southern Nevada. Geol. Soc. Am. Bull. 88, 943–959.

Christie, C. V. L. (1981). "Nuclear Waste Disposal Facility Intrusion: An Archeologists View," Rep. LA-UR-81-830. Los Alamos Natl. Lab., Los Alamos, New Mexico.

Claiborne, H. C. *et al.* (1980). "Expected Environment in High-level Nuclear Waste and Spent Fuel Repositories in Salt," Rep. ORNL/TM-7201. Oak Ridge Natl. Lab., Oak Ridge, Tennessee.

Clark, J. A. (1980). "The reconstruction of the Laurentide Ice Sheet of North America from sea level data: Method and preliminary results. *JGR, J. Geophys. Res.* 85, 4307–4323.

Cleveland, J. M. and Rees, T. F. (1981). Characterization of plutonium in Maxey Flats radioactive trench leachates. *Science* 212, 1506–1509.

Coates, D. R. (1981). "Environmental Geology." Wiley, New York.

Cohen, B. L. (1977). The disposal of radioactive wastes from fission reactors. *Sci. Am.* 236, 21–31.

Cohen, J. J. and King, W. C. (1978). "Determination of a Radioactive Waste Classification System," Rep. UCRL-52535. Lawrence Livermore Lab., Livermore, California.

Colglazier, E. W., ed. (1982). "The Politics of Nuclear Waste." Pergamon, Oxford.

Colombo, P. *et al.* (1979). "Analysis and Evaluation of a Radioactive Waste Package Retrieved from the Atlantic 2800 Meter Disposal Site," Rep. BNL-51102. Brookhaven Natl. Lab., Upton, New York.

Committee on Marine Geosciences (1983). *Whither the Ocean Geosciences?* CMG, Int. Union Geol. Sci., Oslo.

Coordinating Group for Geological Disposal (1983). "The Status of Development of Geological Disposal of Radioactive Wastes," Draft Rep. GD/DOC(83)2. CGGD, Org. Econ. Coop. Dev./Nucl. Energy Agency, Paris.

Cope, C. B. *et al.* (1983). "The Scientific Management of Hazardous Wastes." Cambridge Univ. Press, London and New York.

Costa, J. E. and Baker, V. R. (1981). "Surficial Geology. Building with the Earth." Wiley, New York.

Cowan, G. A. (1977). "Migration Paths for Oklo Reactor Products and Applications to the Problem of Geological Storage of Nuclear Wastes," Rep. LA-UR-77-2787. Los Alamos Natl. Lab., Los Alamos, New Mexico.

Crisman, D. P. and Jacobs, G. K. (1982). "Native Copper Deposits of the Portage Lake Volcanics, Michigan: Their Implications with Respect to Canister Stability for Nuclear Waste Isolation in the Columbia River Basalts Beneath the Hanford Site, Washington," Rep. RHO-BW-ST-26P. Rockwell Hanford Operations, Richland, Washington.

Crowe, B. M. (1980). "Disruptive Event Analysis: Volcanism and Igneous Intrusion," Rep. PNL-2882. Pacific Northwest Lab., Richland, Washington.

Crowe, B. M. and Amos, R. (1982). "Aspects of Possible Magmatic Disruption of a High-level Radioactive Waste Repository in Southern Nevada," Rep. LA-9326-MS. Los Alamos Natl. Lab., Los Alamos, New Mexico.

Crowe, B. M. *et al.* (1982). Calculation of the probability of volcanic disruption of a high-level radioactive waste respository within southern Nevada, U.S.A. *Radioact. Waste Manage.* 3, 167–190.

Crowe, B. M. *et al.* (1983). Aspects of potential magmatic disruption of a high-level radioactive waste repository in southern Nevada. *J. Geol.* 91, 259–276.

Curtis, D. B. and Gancarz, A. J. (1983). "Radiolysis in Nature: Evidence from the Oklo Natural Reactors," Tech. Rep. 83–10. Swed. Nucl. Fuel Supply Co., Div. KBS, Stockholm.

D'Alessandro, M. *et al.* (1980). Probability analysis of geological processes: A useful tool for the safety assessment of radioactive waste disposal. *Radioact. Waste Manage.* 1, 25–42.

Damuth, J. E. *et al.* (1983). Sedimentation processes in the Northwest Pacific Basin revealed by echo-character mapping studies. *Geol. Soc. Am. Bull.* 94, 381–395.

Davis, S. N. (1980). "Hydrogeologic Effects of Natural Disruptive Events on Nuclear Waste Repositories," Rep. PNL-2858. Pacific Northwest Lab., Richland, Washington.

Davis, S. N. (1982). "Workshop on Hydrology of Crystalline Basement Rocks," Rep. LA-8912C. Los Alamos Natl. Lab., Los Alamos, New Mexico.

Dawson, P. R. (1978). "Buoyant Movement of Nuclear Waste Canisters in Marine Sediments," Rep. SAND78-0841. Sandia Labs, Albuquerque, New Mexico.

Dean, W. E. and Schreiber, B. C., eds. (1978). "Marine Evaporites," Short Course No. 4. Soc. Econ. Paleontol. Mineral., Tulsa, Oklahoma.

Deep Sea Drilling Project (1983). "Initial Reports of the Deep Sea Drilling Project LXVIII." U.S. Govt. Printing Office, Washington, D. C.

Deese, D. A. (1978). "Nuclear Power and Radioactive Waste." Lexington Books, Lexington, Massachussets.

Deffeyes, K. S. and MacGregor, I. D. (1980). World uranium resources. *Sci. Am.* 242, 50–60.

Delisle, G. (1980). "Berechnungen zur raumzeitlichen Entwicklung des Temperaturfeldes um ein Endlager für mittel- und hochaktive Abfälle in einer Salzformation. *Z. Dtsch. Geol. Ges.* 131, 461–482.

de Marsily, G. (1981). Nuclear waste disposal: Who will make the final decision? *Radioact. Waste Manage.* 2, 109–121.

de Marsily, G. and Merriam, D. E. (1982). "Predictive Geology." Pergamon, Oxford.

de Marsily, G. *et al.* (1977). Nuclear waste disposal. Can the geologist guarantee isolation? *Science* 197, 519–527.

DeMier, W. V. *et al.* (1979). "GETOUT - A computer Program for Predicting Radionuclide Decay Chain Transport through Geologic Media," Rep. PNL-2970. Pacific Northwest Lab., Richland, Washington.

Denton, G. H. and Hughes, T. J., eds. (1981). "The Last Great Ice Sheets." Wiley, New York.

Desaulniers, D. E. *et al.* (1981). Origin, age and movement of pore water in argillaceous Quaternary deposits at four sites in southwestern Ontario. *J. Hydrol.* 50, 231–257.

Dickin, A. (1981). Hydrothermal leaching of rhyolite glass in the environment has implications for nuclear waste disposal. *Nature (London)* 294, 342–347.

Dickinson, W. R. (1972). "Evidence for plate tectonic regimes in the rock record. *Am. J. Sci.* 272, 551–576.

Doctor, P. G. (1979). "Evaluation for Kriging Techniques for High Level Radioactive Waste Repository Site Characterization," Rep. PNL-2903. Pacific Northwest Lab., Richland, Washington.

Doctor, P. G. (1980). The use of geostatistics in high-level radioactive waste repository site characterization. *Radioact. Waste Manage.* 1, 193–210.

Dole, L. R. *et al.* (1983). "Cement-based Radioactive Waste Hosts Formed under Elevated Temperatures and Pressures (FUETAP concretes) for Savannah River Plant High-level Defense Waste," Rep. ORNL/TM-8579. Oak Ridge Natl. Lab., Oak Ridge, Tennessee.

Donath, F. A. and Pohl, R. O. (1982). Debate on radioactive waste disposal. *Phys. Today* 35, 36–45.

Dosch, R. G. (1980). "Assessment of Potential Radionuclide Transport in Site-specific Geologic Formations," Rep. SAND79-2468. Sandia Labs., Albuquerque, New Mexico.

Dougherty, D. and Adams, J. W. (1983). "Evaluation of the Three Mile Island Unit 2 Reactor Building Decontamination Process," Rep. BNL-NUREG-51689. Brookhaven Natl. Lab., Upton, New York.

Drever, J. I. (1982). "The Geochemistry of Natural Waters." Prentice-Hall, Englewood Cliffs, New Jersey.

Eaton, R. R. and Reda, D. C. (1982). The influence of convective energy transfer on calculated temperature distributions in proposed hard-rock nuclear waste repositories. *Radioact. Waste Manage.* **2**, 343–361.

Eaton, R. R. *et al.* (1980). "Sensitivity of Calculated Pore Fluid Pressure to Repository Fracture Geometry," Rep. SAND80-23380. Sandia Labs., Albuquerque, New Mexico.

Eicher, D. L. (1968). "Geologic Time." Prentice-Hall, Englewood Cliffs, New Jersey.

Elder, J. (1976). "The Bowels of the Earth." Oxford Univ. Press, London and New York.

Elder, J. (1981). "Geothermal Systems." Academic Press, London.

Elsam/Elkraft (1981). "Disposal of High-level Waste from Nuclear Power Plants in Denmark; Salt Dome Investigations." Elsam, Fredericia, Denmark and Elkraft, Ballerup, Denmark.

Emiliani, C. (1966). Paleotemperature analysis of Caribbean cores P6304-8 and P6304-9 and a generalized temperature curve for the past 425,000 years. *J. Geol.* **74**, 109–124.

English, T. *et al.* (1977). "An Analysis of the Technical Status of High Level Radioactive Waste and Spent Fuel Management Systems," Rep. 77-69. Jet Propulsion Lab., Pasadena, California.

Essaid, S. and Klar, K. (1982). Zum Innenbau der Salzstruktur Asse. *Z. Dtsch. Geol. Ges.* **133**, 135–154.

Evans, A. M. (1980). "An Introduction to Ore Geology." Blackwell, Oxford.

Everett, L. G. (1980). "Groundwater Monitoring." General Electric Co., Schenectady, New York.

Ewing, R. C. (1976). Metamict mineral alteration; an implication for radioactive waste disposal. *Science* **192**, 1336–1337.

Fairchild, P. D. and Jenks, G. H. (1978). "Avery Island - Dome Salt *in situ* Test," Rep. Y/OWI/TM-55. Office of Waste Isolation, Oak Ridge, Tennessee.

Falconer, K. L. *et al.* (1982). "Site Selection Criteria for Shallow Land Burial of Low-level Radioactive Waste," Rep. CONF-820303-40. Idaho Natl. Eng. Lab., Idaho Falls, Idaho.

Faure, G. (1977). "Principles of Isotope Geology." Wiley, New York.

Fluck, P. *et al.* (1980). The Variscan units east and west of the Rhine graben. *Mem. BRGM* **108**, 112–131.

Fogg, G. E. and Kreitler, C. W. (1981). Ground-water hydrology around salt domes in the East Texas basin: A practical approach to the contaminant transport problem. *Bull. Assoc. Eng. Geol.* **18**, 387–411.

Foley, M. G. *et al.* (1982). "Geologic Simulation Model for a Hypothetical Site in the Columbia Plateau," Vol. 2, Rep. PNL-3542-2. Pacific Northwest Lab., Richland, Washington.

Forex Neptune (1980). "Feasibility Study for Large Diameter Boreholes for the Deep Drilling Concept of a High-level Waste Repository," Tech. Rep. 80-04. National Cooperative for Radioactive Waste Storage, Baden, Switzerland.

Freeze, R. A. and Cherry, J. A. (1979). "Groundwater." Prentice-Hall, Englewood Cliffs, New Jersey.

Fried, S., ed. (1979). "Radioactive Waste in Geologic Storage." Am. Chem. Soc., Washington, D. C.

Fromm, E. *et al.* (1980). "Introduction to the Geology of Sweden," 26th Int. Geol. Congr., Paris.

Fyfe, W. S. (1980). Radioactive waste disposal solutions: A geological perspective. *Episodes* **3**, 19–22.

Fyfe, W. S. *et al.* (1978). "Fluids in the Earth's Crust." Elsevier, Amsterdam.

Gass, I. G. (1982). Ophiolites. *Sci. Am.* **247**, 108–117.

Geological Society of Stockholm (1982). Dating methods and their application within the Quaternary time scale. *Geol. Foeren. Stockholm, Foerh.* **104**, 255–285.

Gera, F. (1972). Review of salt tectonics in relation to the disposal of radioactive wastes in salt formations. *Geol. Soc. Am. Bull.* **83**, 3551–3574.

Gera, F. (1974). The classification of radioactive wastes. *Health Phys.* **27**, 113–121.

Gera, F. and Jacobs, D. G. (1972). "Considerations in the Long-term Management of High-level Radioactive Wastes," Rep. ORNL-4762. Oak Ridge Natl. Lab., Oak Ridge, Tennessee.

Gerlach, S. A. (1981). "Marine Pollution. Diagnosis and Therapy." Springer, New York.

Getzen, R. T. (1977). Analog-model analysis of regional three-dimensional flow in the groundwater reservoir of Long Island, New York. *Geol. Surv. Prof. Pap. (U.S.)* **982**, 1–49.

Giletti, B. *et al.* (1978). "State of Geological Knowledge Regarding Potential Transport of High-level Radioactive Waste from Deep Continental Repositories," Rep. EPA/520/4-78-004. U.S. Environ. Prot. Agency, Washington, D. C.

Gilmore, W. R., ed. (1977). "Radioactive Waste Disposal. Low and High Level." Noyes Data Corp., Park Ridge, New Jersey.

Goetz, A. F. H. and Rowan, L. C. (1981). Geologic remote sensing. *Science* **211**, 781–791.

Gomez, L. S. *et al.* (1980). "Biological Ramifications of the Subseabed Disposal of High-level Nuclear Waste," Rep. SAND79-2117. Sandia Labs., Albuquerque, New Mexico.

Gonzales, S. (1982). Host rocks for radioactive-waste disposal. *Am. Sci.* **70**, 191–200.

Gore, B. F. *et al.* (1981). "Factors Affecting Criticality for Spent-fuel Materials in a Geologic Setting," Rep. PNL-3791. Pacific Northwest Lab., Richland, Washington.

Goudie, A. S. and Pye, K., eds. (1983). "Chemical Sediments and Geomorphology." Academic Press, London.

Graebner, R. *et al.* (1981). Three-dimensional methods in seismic exploration. *Science* **211**, 535–540.

Granath, V. C. H. *et al.* (1983). The Notch Peak contact metamorphic aureole, Utah: Petrology of the Big Horse Limestone Member of the Orr Formation. *Geol. Soc. Am. Bull.* **94**, 889–906.

Grant, M. A. *et al.* (1982). "Geothermal Reservoir Engineering." Academic Press, New York.

Grantham, L. F. *et al.* (1983). "Process Description and Plant Design for Preparing Ceramic High-level Waste Forms," Rep. DOE/ET/41900-16. U.S. Dept. of Energy, Washington, D.C.

Grauer, R. (1983). "Gläser zur Verfestigung von hochaktivem Abfall: ihr Verhalten gegenüber Wässern," Rep. 477. Eidg. Inst. Reaktorforsch., Würenlingen, Switzerland.

Grauer, R. (1984). "Kristalline Stoffe zur Verfestigung hochradioaktiver Abfälle," Rep. 508. Eidg. Inst. Reaktor-forsch. Würenlingen, Switzerland.

Greenkorn, R. A. (1983). "Flow Phenomena in Porous Media." Dekker, Basel.

Gretener, P. E. (1967). Significance of the rare event in geology. *Am. Assoc. Pet. Geol. Bull.* **51**, 2197–2206.

Grim, R. E. and Güven, N. (1978). "Bentonites: Geology, Mineralogy, Properties and Uses." Elsevier, Amsterdam.

Grisak, G. E. and Cherry, J. A. (1975). Hydrologic characteristics and response of fractured till and clay confining a shallow aquifer. *Can. Geotech. J.* **12**, 23–43.

Grover, J. R. (1980). High-level waste solidification—why we chose glass. *Radioact. Waste Manage.* **1**, 1–12.

Gustavson, T. C. *et al.* (1981). Retreat of the caprock escarpment and denudation of the Rolling Plains in the Texas Panhandle. *Bull. Assoc. Eng. Geol.* **18**, 413–422.

Haggerty, S. E. (1983). Radioactive nuclear waste stabilization: Aspects of solid-state engineering and applied geochemistry. *Annu. Rev. Earth Planet. Sci.* **11**, 133–163.

Hardy, M. P. *et al.* (1978). "Numerical Modeling of Rock Stresses Within a Basaltic Nuclear Waste Repository. Phase I. Problem Definition," Rep. RHO-BWI-C-24. Rockwell Hanford Operations, Richland, Washington.

Harland, W. B. *et al.* (1982). "A Geologic Time Scale." Cambridge Univ. Press, London and New York.

Harmon, K. M. *et al.* (1980). "Summary of Non-U.S. National and International Radioactive Waste Management Programs," Rep. PNL-3333. Pacific Northwest Lab., Richland, Washington.

Harries, J. R. (1980). "Deep Circulation in the Indian and Pacific Oceans and its Implication for the Dumping of Low-level Radioactive Waste," Rep. AAEC/E495. Aust. At. Energy Comm., Lucas Heights, Australia.

Heard, H. C. (1982). "Mechanical, Thermal and Fluid Transport Properties of Rock at Depth," Rep. UCRL-87582. Lawrence Livermore Lab., Livermore, California.

Heckmann, R. A. *et al.* (1979). "Site Suitability Criteria for Solidified High Level Waste Repositories," Rep. UCID-17831. Lawrence Livermore Lab., Livermore, California.

Heiken, G. (1979). Pyroclastic flow desposits. *Am. Sci.* 67, 564–571.

Heintz, W. (1980). Endlagerfähige radioaktive Abfälle. *Z. Dtsch. Geol. Ges.* 131, 343–355.

Herrmann, A. G. (1979). Geowissenschaftliche Probleme bei der Endlagerung radioaktiver Substanzen in Salzdiapiren Norddeutschlands. *Geol. Rundsch.* 68, 1076–1106.

Herrmann, A. G. (1983). "Radioaktive Abfälle. Probleme und Verantwortung." Springer-Verlag, Berlin and New York.

Heuze, F. E. (1982). "On the Modeling of Nuclear Waste Disposal by Rock Melting," Rep. UCRL-87510. Lawrence Livermore Lab., Livermore, California.

Heuze, F. E. (1983). High-temperature mechanical, physical and thermal properties of granitic rocks—a review. *Int. J. Rock Mech. Min. Sci.* 20, 3–10.

Heuze, F. E. *et al.* (1982). Rock mechanics studies of mining in the Climax granite. *Int. J. Rock Mech. Min. Sci.* 19, 167–183.

Hill, D. *et al.* (1982). Management of high-level waste repository siting. *Science* 218, 859–864.

Hobson, G. D., ed. (1980). "Developments in Petroleum Geology," Vol. 2. Appl. Sci. Publ., London.

Hofmann, P. L. and Breslin, J. J., eds. (1981). "The Technology of High-Level Nuclear Waste Disposal," Vol. 1. U.S. Dept. of Energy, Washington, D. C.

Hollister, C. D. *et al.* (1981). Seabed disposal of nuclear wastes. *Science* 213, 1321–1326.

Howell, R. E. and Olsen, D. (1982). "Citizen Participation in Nuclear Waste Repository Siting," Rep. ONWI-267. Office of Nuclear Waste Isolation, Columbus, Ohio.

Hsü, K. J., ed. (1982). "Mountain Building Processes." Academic Press, London.

Hsü, K. J. (1983). Actualistic catastrophism. *Sedimentology* 30, 3–9.

Hubbell, D. W. and Glenn, J. J. (1973). Distribution of radionuclides in bottom sediments of the Columbia river estuary. *Geol. Surv. Prof. Pap. (U.S.)* 433-L, 1–63.

Huff, D. D. *et al.* (1983). Hydrologic factors and ^{90}Sr transport: a case study. *Environ. Geol.* 4, 53–63.

Imbrie, J. and Imbrie, K. P. (1979). "Ice Ages. Solving the Mystery." Enslow, Short Hills, New Jersey.

International Atomic Energy Agency (IAEA) (1974), "Formation of Uranium Ore Deposits." IAEA, Vienna.

International Atomic Energy Agency (IAEA) (1975). "The Oklo Phenomenon." IAEA, Vienna.

International Atomic Energy Agency (IAEA) (1977). "Site Selection Factors for Repositories of Solid High-level and Alpha-bearing Wastes in Geological Formations." Tech. Rep. 177. IAEA, Vienna.

International Atomic Energy Agency (IAEA) (1978a). "The Oceanographic Basis of the IAEA Revised Definition and Recommendations Concerning High-level Radioactive Waste Unsuitable for Dumping at Sea." Tech. Doc. 210. IAEA, Vienna.

International Atomic Energy Agency (IAEA) (1978b). "The Radiological Basis of the IAEA Revised Definition and Recommendations Concerning High-level Radioactive Waste Unsuitable for Dumping at Sea." Tech. Doc. 211. IAEA, Vienna.

International Atomic Energy Agency (IAEA) (1980a). "International Nuclear Fuel Cycle Evaluation." IAEA, Vienna.

International Atomic Energy Agency (IAEA) (1980b). "Underground Disposal of Radioactive Wastes." IAEA, Vienna.

International Atomic Energy Agency (IAEA) (1981). "Management of Alpha-Contaminated Wastes." IAEA, Vienna.

International Atomic Energy Agency (IAEA) (1982a). "Management of Wastes from Uranium Mining and Milling." IAEA, Vienna.

International Atomic Energy Agency (IAEA) (1982b). "Environmental Migration of Long-Lived Radionuclides." IAEA, Vienna.

International Atomic Energy Agency (IAEA) (1982c). "Vein-Type and similar Uranium Deposits in Rocks Younger than Proterozoic." IAEA, Vienna.

International Atomic Energy Agency (IAEA) (1983a). "Decommissioning of Nuclear Facilities: Decontamination, Dissassembly and Waste Management." Tech. Rep. 230. IAEA, Vienna.

International Atomic Energy Agency (IAEA) (1983b). "Disposal of Radioactive Grouts into Hydraulically Fractured Shale." Tech. Rep. 232. IAEA, Vienna.

International Atomic Energy Agency (IAEA) (1983c). "Conditioning of Radioactive Wastes for Storage and Disposal." IAEA, Vienna.

International Atomic Energy Agency (IAEA) (1983d). "Control of Radioactive Waste Disposal into the Marine Environment." Saf. Ser. 61. IAEA, Vienna.

International Atomic Energy Agency (IAEA) (1983e). "Isotope Techniques in the Hydrogeological Assessment of Potential Sites for the Disposal of High-level Radioactive Wastes." Tech. Rep. 228. IAEA, Vienna.

International Atomic Engery Agency (IAEA) (1983g). "Nuclear Power Experience. Vol. 3, Nuclear Fuel Cycle." IAEA, Vienna.

International Atomic Energy Agency (IAEA) (1983g). "Nuclear Power Experience," Vol. 3. IAEA, Vienna.

Jackson, T. C., ed. (1981). "Nuclear Waste Management. The Ocean Alternative." Pergamon, Oxford.

Jensen, B. S. (1982). "Migration Phenomena of Radionuclides into the Geosphere." Harwood Academic Publ., Chur, Switzerland.

Jet Propulsion Laboratory (1978). "Comparative Assessment of Terrestrial Options for Nuclear Waste Disposal," Rep. 5030-170. JPL, Pasadena, California.

Johansson, T. B. and Steen, P. (1978). "Radioactive Waste from Nuclear Power Plants: Facing the Ringhals-3 Decision," Rep. Ds I 1978. Swedish Dept. of Industry, Stockholm.

Johnpeer, G. D. et al. (1981). "Assessment of Potential Volcanic Hazards: Pasco Basin, Washington," Rep. RHO-BW-CR-130P. Rockwell Hanford Operations, Richland, Washington.

Johnston, R. J. and Palmer, R. A. (1982). "Stability of Ceramic Waste Forms in Potential Repository Environments: A Review," Rep. DOE/ET/41900-13. U.S. Dept. of Energy, Washington, D.C.

Johnston, R. M. (1983). "The Conversion of Smectite to Illite in Hydrothermal Systems: A Literature Review," Rep. AECL-7792. Atomic Energy of Canada Ltd., Whiteshell, Manitoba.

Johnstone, J. K. and Wolfsberg, K. (1980). "Evaluation of Tuff as a medium for a Nuclear Waste Repository: Interim Status Report on the Properties of Tuff," Rep. SAND80-1464. Sandia Labs., Albuquerque, New Mexico.

Just, R. A. (1978). "Heat Transfer Studies in Salt and Granite," Rep. ORNL/ENG/TM-14. Oak Ridge Natl. Lab., Oak Ridge, Tennessee.

Karkhanis, S. N. et al. (1980). Leaching behaviour of rhyolite glass. Nature (London) 284, 435–437.

Kaser, J. D. et al. (1976). "Fusion Reactor Radioactive Waste Management," Rep. ARH-SA-275. Atlantic Richfield Hanford Co., Richland, Washington.

Kasper, R. B. (1981). "Field Study of Plutonium Transport in the Vadose Zone." Rep. RHO-SA-224. Rockland Hanford Operations, Richland, Washington.

Kaufmann, A. M. et al. (1978). "Calculated Radionuclide Migration from a Geologic High Level Waste Repository," Rep. UCRL-81148. Lawrence Livermore Lab., Livermore, California.

Kelly, J. E. and Shea, C. E. (1982). The Subseabed Disposal Program for high-level radioactive waste—public response. Oceanus 25, 42–53.

Kennett, J. P. (1982). "Marine Geology." Prentice-Hall, Englewood Cliffs, New Jersey.

Kesson, S. E. and Ringwood, A. E. (1983). Safe disposal of spent nuclear fuel. Radioact. Waste Manage. 4, 159–174.

Kibbey, A. H. and Godbee, H. W. (1980). "State-of-the-Art Report on Low-level Radioactive Waste Treatment," Rep. ORNL/TM-7427. Oak Ridge Natl. Lab., Oak Ridge, Tennessee.

Klingsberg, C. and Duguid, J. (1982). Isolating radioactive wastes. Am. Sci. 70, 182–190.

Kocher, D. C. et al. (1983). "Uncertainties in Geologic Disposals of High-level Wastes: Groundwater Transport of Radionuclides and Radiological Consequences," Rep. ORNL-5838. Oak Ridge Natl. Lab., Oak Ridge, Tennessee.

Kovari, K., ed. (1983). "Field Measurements in Geomechanics." Balkema, Rotterdam.

Kranz, R. L. (1980). The effects of confining pressure and stress difference on static fatigue of granite. JGR, J. Geophys. Res. 85, 1854–1866.

Kreiter, M. et al. (1977). "Influence of Plutonium Recycle on the Radioactive Waste Management System," Rep. BNWL-SA-6193. Pacific Northwest Lab., Richland, Washington.

Kresten, P. and Chyssler, J. (1976). The Götemar massif in southeastern Sweden: A reconnaissance study. Geol. Foeren. Stockholm Foerh. 98, 155–161.

Kukla, G. J. (1977). Pleistocene land-sea correlations. I. Europe. Earth Sci. Rev. 13, 307–374.

Landa, E. (1980). Isolation of uranium mill tailings and their component radionuclides from the biosphere: Some earth science perspectives. Geol. Surv. Circ. (U.S.) 814, 1–32.

Landström, O. et al. (1983). "Migration Experiments at Studsvik," Tech. Rep. 83–18. Swedish Nucl. Fuel Supply Co., Div. KBS, Stockhom.

Lappin, A. R. (1980). "Thermal Conductivity of Silicic Tuffs: Predictive Formalism and Comparison with Preliminary Experimental Results," Rep. SAND80-0769. Sandia Labs., Albuquerque, New Mexico.

Laronne, J. B. and Moseley, M. P., eds. (1982). "Erosion and Sediment Yield." Hutchinson Ross, Stroudsburg, Pennsylvania.

La Sala, A. M. and Doty, G. C. (1971). Preliminary evaluation of hydrologic factors related to waste storage in basaltic rocks at the Hanford Reservation, Washington. Geol. Surv. (U.S.) Open-File Rep.

Last, G. V. (1983). "Radionuclide Distributions around a Low-level Radioactive Waste Disposal Pond and Ditch System at the Hanford Site." Rep. RHO-HS-SA-19-P. Rockwell Hanford Operations, Richland, Washington.

Lawrence Berkeley Laboratory (1978). "Geotechnical Assessment and Instrumentation Needs for Nuclear Waste Isolation in Crystalline and Argillaceous Rock," Rep. LBL-7096. LBL, Berkeley, California.

Lawrence Livermore Laboratory (1978). "Final Report on Uncertainties in the Detection,

Measurement and Analysis of Selected Features Pertinent to Deep Geologic Repositories," Rep. UCRL-13912. LLL, Livermore, California.

Lawrence Livermore Laboratory (1979). "Preliminary Study of Radioactive Waste Disposal in the Vadose Zone," Rep. UCRL-13992. LLL, Livermore, California.

Lawrence Livermore Laboratory (1980a), "State-of-the-Art for Evaluating the Potential Effects of Erosion and Deposition on a Radioactive Waste Repository," Rep. UCRL-15268. LLL, Livermore, California.

Lawrence Livermore Laboratory (1980b). "State-of-the-Art for Evaluating the Potential Impact of Tectonism and Volcanism on a Radioactive Waste Repository," Rep. UCRL-15269. LLL, Livermore, California.

Lawrence Livermore Laboratory (1980c). "State-of-the-Art for Evaluating the Potential Impact of Flooding on a Radioactive Waste Repository," Rep. UCRL-15270. LLL, Livermore, California.

Lawrence Livermore Laboratory (1982). "Nuclear Waste Storage: Evaluation the Uncertainties," Energy Technol. Rev., May 1982, pp. 22–33. LLL, Livermore, California.

LeAnderson, J. et al. (1983). Origin of fluids and metals in porphyry and epithermal mineral deposits. Geology 11, 557–558.

Ledbetter, J. O. et al. (1975). "Radioactive Waste Management by Burial in Salt Domes," Rep. EMRL 1112. Eng. Mech. Res. Lab., University of Texas, Austin.

Leeder, M. R. (1982). "Sedimentology. Process and Product." Allen & Unwin, London.

Lehman, J. P., ed. (1983). "Hazardous Waste Disposal." Plenum, New York.

Le Pichon, X. et al. (1973). "Plate Tectonics." Elsevier, Amsterdam.

Lindblom, U. E. and Gnirk, P. F. (1982). "Nuclear Waste Disposal. Can We Rely on Bedrock?" Pergamon, Oxford.

Lindsey, C. G. et al. (1983). "Survivability of Ancient Man-made Earthern Mounds: Implications for Uranium Mill Tailings Impoundments," Rep. NUREG/CR-3061. Pacific Northwest Lab., Richland, Washington.

Lipschutz, R. D. (1980). "Radioactive Waste: Politics, Technology and Risk." Ballinger, Cambridge, Massachusetts.

Llewellyn, G. H. (1978). "Prediction of Temperature Increases in a Salt Repository Expected from the Storage of Spent Fuel or High-level Waste," Rep. ORNL/ENG/TM-7. Oak Ridge Natl. Lab., Oak Ridge, Tennessee.

Lofgren, G. (1971). Experimentally produced devitrification textures in natural rhyolitic glass. Geol. Soc. Am. Bull. 82, 111–123.

Lutze, W., ed. (1982). "Scientific Basis for Nuclear Waste Management," Vol. V. North-Holland, New York.

McAlester, A. L. (1968). "The History of Life." Prentice-Hall, Englewood Cliffs, New Jersey.

Macbeth, P. J. et al. (1978). "Screening of Alternative Methods for the Disposal of Low-level Radioactive Wastes," Rep. NUREG/CR-0308. U.S. Nuc. Regul. Comm., Washington, D.C.

McCall, P. L. and Tevesz, M. J. S., eds. (1982). "Animal-Sediment Relations." Plenum, New York.

McCarthy, G. J., ed. (1979). "Scientific Basis for Nuclear Waste Management," Vol. I. Plenum, New York.

McCarthy, G. J. et al. (1978). Interactions between nuclear waste and surrounding rock. Nature (London) 273, 216–217.

Macdonald, G. (1972). "Volcanoes." Prentice-Hall, Englewood Cliffs, New Jersey.

McElhinny, M. W., ed. (1979). "The Earth: Its Origin, Structure and Evolution." Academic Press, London.

McLaren, D. J. (1983). Bolides and biostratigraphy. Geol. Soc. Am. Bull. 94, 313–324.

McVey, D. F. *et al.* (1979). "Thermal, Chemical and Mass-Transport Processes Induced in Abysall Sediments by the Emplacement of Nuclear Waste: Experimental and Modeling Results," Rep. SAND79-2244C. Sandia Labs., Albuquerque, New Mexico.

Mahaffy, M. A. (1976). A three-dimensional numerical model of ice sheets - tests on the Barnes Ice Cap, Northwest Territories. *JGR, J. Geophys. Res.* **81**, 1059–1066.

Mair, J. A. and Green, A. G. (1981). High-resolution seismic reflection profiles reveal fracture zones within a homogeneous granite batholith. *Nature (London)* **294**, 439–442.

Majer, E. L. *et al.* (1981). Monitoring an underground repository with modern seismological methods. *Int. J. Rock Mech. Min. Sci.* **18**, 517–527.

Mallio, W. J. and Peck, J. H. (1981). Practical aspects of geological prediction. *Bull. Assoc. Eng. Geol.* **18**, 369–374.

Marine, I. W. (1979). "Hydrology of Buried Crystalline Rock at Savannah River Plant near Aiken, South Carolina," Rep. DOE/SR-WM-79-2. U.S. Dept. of Energy, Savannah River Operations Office, Aiken, South Carolina.

Martinez, J. D. and Thomas, R. L., eds. (1977). "Salt Dome Utilization and Environmental Considerations." Inst. Environ. Stud., Louisiana State University, Baton Rouge.

Martinez, J. D. *et al.* (1975). "Investigation of the Utility of Gulf Coast Salt Domes for the Storage of Radioactive Wastes," Rep. ORNL/sub-4112/10. Oak Ridge Natl. Lab., Oak Ridge, Tennessee.

Marvin, U. B. (1973). "Continental Drift. Evolution of a Concept." Smithson. Inst. Press, Washington, D.C.

Mauthe, F. (1979). Probleme und Risiken bei der geplanten Einlagerung radioaktiver Abfälle in einen nordwestdeutschen Salzstock. *Mitt. Geol. Inst. Hannover* **18**, 1–60.

Mayer-Rosa, D. and Cadiot, B. (1979). A review of the 1356 Basel earthquake: Basic data. *Tectonophysics* **53**, 325–333.

Maynard, J. B. (1983). "Geochemistry of Sedimentary Ore Deposits." Springer, New York.

Means, J. L. *et al.* (1978). Migration of radioactive wastes: Radionuclide mobilization by complexing agents. *Science* **200**, 1477–1481.

Melnyk, T. W. *et al.* (1983). "High-level Waste Glass Field Burial Tests at CRNL: The Effect of Geochemical Kinetics on the Release and Migration of Fission Products in a Sandy Aquifer," Rep. AECL-6836. Atomic Energy of Canada Ltd., Whiteshell, Manitoba.

Mercier, J. W. *et al.* (1983). "Role of the Unsaturated Zone in Radioactive and Hazardous Waste Disposal." Ann Arbor Sci. Publ., Ann Arbor, Michigan.

Merewether, E. A. *et al.* (1973). Shale, mudstone and claystone as potential host rocks for underground emplacement of waste. *Geol. Surv. Open-File Rep. (U.S.).* 1–90.

Merrett, G. J. and Gillespie, P. A. (1983). "Nuclear Fuel Waste Disposal: Long-term Stability Analysis," Rep. AECL-6820. Atomic Energy of Canada Ltd., Whiteshell, Manitoba.

Milnes, A. G. (1979). Swiss go to polls again on nuclear issue. *Nature (London)* **278**, 500–501.

Milnes, A. G. *et al.* (1980). Endlagerungskonzepte für radioaktive Abfälle im Ueberblick. *Z. Dtsch. Geol. Ges.* **131**, 359–385.

Moore, J. G., ed. (1981). "Scientific Basis for Nuclear Waste Management," Vol. III. Plenum, New York.

Morell, D. and Magorian, C. (1982). "Siting Hazardous Waste Facilities." Ballinger, Cambridge, Massachusetts.

Moreno, L. *et al.* (1983). "Evaluation of Some Tracer Tests in the Ganitic Rock at Finnsjön," Tech. Rep. 83-38. Swed. Nucl. Supply Co., Div. KBS, Stockholm.

Mörner, N.-A. (1977). Faulting, fracturing and seismic activity as a function of glacial-isostacy in Fennoscandia. *Geology* **6**, 41–45.

Mörner, N.-A., ed. (1980). "Earth Rheology, Isostacy and Eustasy." Wiley, New York.

Murray, C. N. (1981). Marine radiological studies related to Seabed disposal. *Radioact. Waste Manage.* 2, 83–100.

Murray, C. N. and Stanners, D. A. (1982). Development of an assessment methodology for the disposal of high-level radioactive waste into deep ocean sediments. *Radioact. Waste Manage.* 2, 239–294.

Narasimhan, T. N. (1980). "Groundwater Modeling in Subsurface Nuclear Waste Disposal: An Overview," Rep. LBL-10686, pp. 144–145. Lawrence Berkeley Lab., Berkeley, California.

Narasimhan, T. N., ed. (1982). "Recent Trends in Hydrology," Geol. Soc. Am., Spec. Pap. 189, Geol. Soc. Am., Washington, D. C.

Narasimhan, T. N. *et al.* (1980). The role of pore pressure in deformation in geologic processes. *Geology* 8, 349–351.

National Academy of Sciences (NAS) (1976). "The Shallow Land Burial of Low-Level Radioactively Contaminated Solid Waste." Natl. Acad. Sci., Washington, D. C.

National Academy of Sciences (NAS) (1978). "Radioactive Wastes at the Hanford Reservation. A Technical Review." Natl. Acad. Sci., Washington, D. C.

National Academy of Sciences (NAS) (1983). "A Study of the Isolation System for Geologic Disposal of Radioactive Wastes." Natl. Acad. Sci., Washington, D. C.

National Cooperative for Radioactive Waste Storage (NAGRA) *et al.* (1978). "Die nukleare Entsorgung in der Schweiz." NAGRA, Baden, Switzerland.

National Cooperative for Radioactive Waste Storage (NAGRA) (1981). "Die Endlagerung schwach- und mittel-radioaktiver Abfälle in der Schweiz," Tech. Rep. 81–04. NAGRA, Baden, Switzerland.

National Cooperative for Radioactive Waste Storage (NAGRA) (1982). "Sondierbohrung Weiach: Arbeitsprogramm," Tech. Rep. 82–10. NAGRA, Baden, Switzerland.

National Cooperative for Radioactive Waste Storage (NAGRA) (1983a). "Nucleare Entsorgung Schweiz. Konzept und Stand der Arbeiten 1982," Tech. Rep. 83–02. NAGRA, Baden, Switzerland.

National Cooperative for Radioactive Waste Storage (NAGRA) (1983b). "Wege zum Nachweis der Sicherheit von Endlagern," Tech. Rep. 83–03. NAGRA, Baden, Switzerland.

Neiheisel, J. (1979). "Sediment Characteristics of the 2800 Meter Atlantic Nuclear Waste Disposal Site: Radionuclide Retention Potential," Rep. ORP/TAD-79-10. U.S. Environ. Prot. Agency, Washington, D. C.

Nelson, P. H. and Rachiele, R. (1982). Water migration induced by thermal loading of a granitic rock mass. *Int. J. Rock Mech. Min. Sci.* 19, 353–359.

Neretnieks, I. (1980). Diffusion in the rock matrix: An important factor in radionuclide retardation? *JGR, J. Geophys. Res.* 85, 4379–4397.

Neretnieks, I. (1981). Age dating of groundwater in fissured rock: Influence of water volume in micropores. *Water Resour. Res.* 17, 421.

Nilsson, T. (1983). "The Pleistocene. Geology and Life in the Quaternary Ice Age." Enke Verlag, Stuttgart.

Nishita, H. (1979). "A Review of Behavior of Plutonium in Soils and Other Geologic Materials," Rep. NUREG/CR-1056. U.S. Nucl. Regul. Comm., Washington, D.C.

Northrup, C. J. M., ed. (1980). "Scientific Basis for Nuclear Waste Management," Vol. II. Plenum, New York.

Norton, D. and Knight, J. (1977). Transport phenomena in hydrothermal systems: Cooling plutons. *Am. J. Sci.* 277, 937–981.

Norton, D. and Taylor, H. P. (1979). Quantitative simulation of the hydrothermal systems of crystallizing magmas on the basis of transport theory and oxygen isotope data: an analysis of the Skaergaard intrusion. *J. Petrology* 20, 421–486.

O'Brien, M. T. *et al.* (1979). "The Very Deep Hole Concept: Evaluation of an Alternative for Nuclear Waste Disposal," Rep. LBL-7089. Lawrence Berkeley Lab., Berkeley, California.

Office of Waste Isolation (1976). "The Selection and Evaluation of Thermal Criteria for a Geologic Waste Isolation Facility in Salt," Rep. Y/OWI/SUB-76/07220. OWI, Oak Ridge, Tennessee.

Olkiewicz, A. *et al.* (1979). "Geology and Fracture System at Stripa," Rep. LBL-8907. Lawrence Berkeley Lab., Berkeley, California.

Ollier, C. D. (1981). "Tectonics and Landforms." Longman, London.

Olsen, C. R. *et al.* (1983). "Chemical, Geological and Hydrological Factors Governing Radionuclide Migration from a Formerly Used Seepage Trench: A Field Study," Rep. ORNL/TM-8839. Oak Ridge Natl. Lab., Oak Ridge, Tennessee.

Organization for Economic Cooperation and Development (OECD) (1977). "Objectives, Concepts and Strategies for the Management of Radioactive Waste Arising from Nuclear Power Programmes." OECD, Paris.

Organization for Economic Cooperation and Development (OECD) (1978). "Management, Stabilization and Environmental Impact of Uranium Mill Tailings." OECD, Paris.

Organization for Economic Cooperation and Development (OECD) (1979a). "Guidelines for Sea Dumping Packages of Radioactive Waste." OECD, Paris. 32.

Organization for Economic Cooperation and Development (OECD) (1979b). "The Migration of Long-Lived Radionuclides in the Geosphere." OECD, Paris.

Organization for Economic Cooperation and Development (OECD) (1979c). "Low-Flow, Low-Permeability Measurements in Largely Impermeable Rocks." OECD, Paris.

Organization for Economic Cooperation and Development (OECD) (1980a). "Review of the Continued Suitability of the Dumping Site for Radioactive Waste in the Northeast Atlantic." OECD, Paris.

Organization for Economic Cooperation and Development (OECD) (1980b). "Use of Argillaceous Materials for the Isolation of Radioactive Waste." OECD, Paris.

Organization for Economic Cooperation and Development (OECD) (1981a). "Near-Field Phenomena in Geologic Repositories for Radioactive Wastes." OECD, Paris.

Organization for Economic Cooperation and Development (OECD) (1981b). "Radionuclide Release Scenarios for Geologic Repositories." OECD, Paris.

Organization for Economic Cooperation and Development (OECD) (1981c). "Siting of Radioactive Waste Repositories in Geological Formations." OECD, Paris.

Organization for Economic Cooperation and Development (OECD) (1982a). "Disposal of Radioactive Waste. An Overview of the Principles Involved." OECD, Paris.

Organization for Economic Cooperation and Development (OECD) (1982b). "Geological Disposal of Radioactive Waste. Research in the OECD Area." OECD, Paris.

Organization for Economic Cooperation and Development (OECD) (1982c). "Geological Disposal of Radioactive Waste. Geochemical Processes." OECD, Paris.

Organization for Economic Cooperation and Development (OECD) (1983a). "Etat des connaissances océanographiques relatives au site d'immersion de déchets radioactifs de faible activité dans l'Atlantique Nord-Est." OECD, Paris.

Organization for Economic Cooperation and Development (OECD) (1983b). "The International Stripa Project." OECD, Paris.

Organization for Economic Cooperation and Development (OECD) (1983c). "In Situ Experiments in Granite." OECD, Paris.

Oversby, V. M. and Ringwood, A. E. (1981). Lead isotopic studies of zirconolite and perovskite and their implications for long range Synroc stability. *Radioact. Waste Manage.* **1**, 289–307.

Owen, H. G. (1983). "Atlas of Continental Displacement." Cambridge Univ. Press, London and New York.

Paige, H. W. *et al.* (1981). Assessment of national systems for obtaining local acceptance of waste management siting and routing activities. *Radioact. Waste Manage.* **2**, 1–48.

Park, C. F. and MacDiarmid, R. A. (1975). "Ore Deposits," 3rd ed. Freeman, San Francisco, California.

Paterson, M. S. (1978). "Experimental Rock Deformation: The Brittle Field." Springer-Verlag, Berlin and New York.

Peltier, W. R. (1981). Ice age geodynamics. *Annu. Rev. Earth Planet. Sci.* **9**, 199–226.

Perkins, B. L. (1982). "Disposal of Liquid Radioactive Wastes Through Wells or Shafts," Rep. LA-9142-MS. Los Alamos Natl. Lab., Los Alamos, New Mexico.

Perry, E. C. and Montgomery, C. W., eds. (1982). "Isotope Studies in Hydrologic Processes." Northern Illinois Univ. Press, DeKalb.

Petrie, G. M. *et al.* (1981). "Geologic Simulation Model for a Hypothetical Site in the Columbia Plateau," Rep. PNL-3542. Pacific Northwest Lab. Richland, Washington.

Philberth, K. (1977). The disposal of radioactive waste in ice sheets. *J. Glaciol.* **19**, 607–617.

Phillips, J. *et al.* (1979). "A Waste Inventory Report for Reactor and Fuel-fabrication Facility Wastes," Rep. ONWI-20-NUS-3314. Office of Nuclear Waste Isolation, Columbus, Ohio.

Pigford, T. H. (1983). "Geologic Disposal of Radioactive Waste - 1983," Rep. LBL-16795. Lawrence Berkeley Lab., Berkeley, California.

Pinnecker, E. V., ed. (1983). "General Hydrology." Cambridge Univ. Press, London and New York.

Potter, G. L. (1978). "Summary of Climatic Input for Waste Management Site Suitability Criteria and State of Progress," Rep. UCID-17904. Lawrence Livermore Lab., Livermore, California.

Powers, D. W. *et al.* (1978). "Geological Characterization Report: Waste Isolation Pilot Plant (W.I.P.P.) Site, Southeastern New Mexico," Rep. SAND-78-1596. Sandia Labs., Albuquerque, New Mexico.

Press, F. and Siever, R. (1983). "Earth," 3rd ed. Freeman, San Francisco, California.

Price, N. J. (1984). "Fault and Joint Development in Brittle and Semi-Brittle Rocks," 2nd ed. Pergamon, Oxford.

Price, S. M. and Ames, L. L. (1975). "Characterization of Actinide-bearing Sediments Underlying Liquid Waste Disposal Facilities at Hanford," Rep. ARH-SA-232. Atlantic Richfield Hanford Co., Richland, Washington.

Ramberg, H. (1981). "Gravity, Deformation and the Earth's Crust," 2nd ed. Academic Press, London.

Ramirez, A. L. (1983). "Borehole-to-Borehole Geophysical Methods Applied to Investigations of High-level Waste Repository Sites," Rep. UCRL-88648. Lawrence Livermore Lab., Livermore, California.

Rasmuson, A. and Neretnieks, I. (1981). Migration of radionuclides in fissured rock: The influence of micropore diffusion and longitudinal dispersion. *JGR, J. Geophys. Res.* **86**, 3749–3758.

Rasmussen, N. H. and Rohay, A. C. (1982). "The Impact of Seismicity on the Stability of an Underground Repository," Rep. RHO-BW-SA-269-P. Rockwell Hanford Operations, Richland, Washington.

Reading, H. G., ed. (1978). "Sedimentary Environments and Facies." Blackwell, Oxford.

Reidel, S. P. (1983). Stratigraphy and petrogenesis of the Grande Ronde Basalt from the deep canyon country of Washington, Oregon and Idaho. *Geol. Soc. Am. Bull.* **94**, 519–542.

Réatháti, L. (1983). "Groundwater in Civil Engineering." Elsevier, Amsterdam.

Revelle, R. (1982). Carbon dioxide and world climate. *Sci. Am.*, **247**, 33–41.

Ringwood, A. E. (1980). Safe disposal of high-level radioactive wastes. *Fortschr. Mineral.* **58**, 149–168.

Ringwood, A. E. (1982). Immobilization of radioactive wastes in SYNROC. *Am. Sci.* **70**, 201–207.

Ringwood, A. E. *et al.* (1979). Immobilization of high-level nuclear reactor wastes in SYN-ROC. *Nature (London)* **278**, 219–223.

Roberts, L. (1979). Radioactive waste—policy and perspective. *Atom* **267**, 8–20.

Roberts, W. H. and Cordell, R. J., eds. (1980). "Problems in Petroleum Migration. Studies in Geology," No. 10. Am. Assoc. Pet. Geol., Tulsa, Oklahoma.

Robertson, J. B. and Fischer, J. (1983). "Geohydrologic Problems at Low-level Radioactive Disposal Sites in the United States of America," Prepr. Pap. IAEA-CN-43/183. Int. At. Energy Agency, Vienna.

Robin, G. de Q. (1975). The disposal of radioactive wastes in the Antarctic ice sheet. *Polar Rec.* **17**, 578–579.

Robin, G. de Q., ed. (1983). "The Climatic Record in the Polar Ice Sheets." Cambridge Univ. Press, London and New York.

Robinson, G. R. *et al.* (1981). Feasibility study of data for nuclear waste disposal: Thermodynamic properties of basalt, granite, shale, and tuff. *Geol. Surv. Open-File Rep. (U.S.)* **81-0470**, 1–135.

Rockwell Hanford Operations (1977). "Basalt Waste Isolation Program Annual Report," Rep. RHO-ST-9. RHO, Richland, Washington.

Roedder, E. and Bassett, R. L. (1981). Problems of determination of the water content of rock-salt samples and its significance in nuclear-waste storage siting. *Geology* **9**, 525–530.

Roedder, E. and Chou, I.-M. (1982). A critique of "Brine migration in salt and its implications in the geologic disposal of nuclear waste" (ORNL 5818). *Geol. Surv. Open-File Rep. (U.S.)* **82-1131**, 1–33.

Rogers, T. H. and Conwell, F. R. (1981). Geologic nuclear waste repository site selection studies in the Paradox Basin, Utah. *Bull. Assoc. Eng. Geol.* **18**, 437–443.

Romer, R. H. (1976). "Energy. An Introduction to Physics." Freeman, San Francisco, California.

Rosenheim, J. S. and Bennett, G. D., eds. (1984). "Groundwater Hydraulics." Amer. Geophys. Union, Washington, D. C.

Ross, B. *et al.* (1979). "NUTRAN: A Computer Model of Long-term Hazards from Waste Repositories," Rep. UCRL-15150. Lawrence Livermore Lab., Livermore, California.

Ross, B. *et al.* (1981). A computer model of long-term hazards from waste repositories. *Radioact. Waste Manage.* **1**, 325–336.

Roy, R. (1982). "Radioactive Waste Disposal," Vol. 1. Pergamon, Oxford.

Runchal, A. and Maini, T. (1980). The impact of a high level nuclear waste repository on the regional ground water flow. *Int. J. Rock Mech. Min. Sci.* **17**, 253–264.

Rutter, E. (1972). Influence of interstitial water on the rheological behaviour of calcite rocks. *Tectonophysics* **14**, 13–33.

Rybach, L. (1975). Thermische Fragen der Lagerung radioaktiven Abfällen in Anhydrit. *Bull. Ver. Schweiz. Pet.-Geol. -Ing.* **41**, 1–13.

Rybach, L. *et al.* (1980). The Swiss Geotraverse Basel–Chiasso—a review. *Eclogae Geol. Helvetiae* **73**, 437–462.

Rydberg, J. (1981). "Groundwater Chemistry of a Nuclear Waste Repository in Granite Bedrock," Rep. UCRL-53155. Lawrence Livermore Lab., Livermore, California.

Sägesser, R. and Mayer-Rosa, D. (1978). Erdbebengefährdung in der Schweiz. *Schweiz. Bauztg.* **96**, 107–127.

Sanders, I. and Young, A. (1983). Rates of surface processes on slopes, slope retreat, and denudation. *Earth Surf. Proc. Landforms* **8**, 473–501.

Sanford, R. F. (1982). Preliminary model of regional Mesozoic groundwater flow and uranium deposition in the Colorado Plateau. *Geology* **10**, 348–352.

Sax, N. I., ed. (1979). "Dangerous Properties of Industrial Materials," 5th ed. Van Nostrand-Reinhold, Princeton, New Jersey.

Schmidt, B. *et al.* (1980). "Thermal and Mechanical Properties of Hanford Basalts: Compilation and Analysis," Rep. RHO-BWI-C-90. Rockwell Hanford Operations, Richland, Washington.

Schubert, G. and Yuen, D. A. (1982). Initiation of ice ages by creep instability and surging of the East Antarctic ice sheet. *Nature (London)* **296**, 127–130.

Schwartz, L. L. *et al.* (1976). "High Level Radioactive Waste Isolation by Incorporation in Silicate Rock," Rep. UCRL-78746. Lawrence Livermore Lab., Livermore, California.

Shackleton, N. J. *et al.* (1984). Oxygen isotope calibration of the onset of ice-rafting and history of glaciation in the North Atlantic region. *Nature (London)* **307**, 620–623.

Sheets, P. D. and Grayson, D. K., eds. (1979). "Volcanic Activity and Human Ecology." Academic Press, New York.

Shephard, L. E., ed. (1983). "1982 Subseabed Disposal Program Annual Report: Site Assessment October 1981 to September 1982," Rep. SAND82-2711. Sandia Labs., Albuquerque, New Mexico.

Sholkovitz, E. R. (1983). The geochemistry of plutonium in fresh and marine water environments. *Earth Sci. Rev.* **19**, 95–161.

Sibson, R. H. (1977). Fault rocks and fault mechanisms. *J. Geol. Soc. London* **133**, 191–213.

Sibson, R. H. (1981). Fluid flow accompanying faulting; field evidence and models. *In* "Earthquake Prediction: An International Review" (D. W. Simpson, ed.), pp. 593–603. Am. Geophys. Union, Washington, D.C.

Sibson, R. H. *et al.* (1975). Seismic pumping: A hydrothermal fluid transport mechanism. *J. Geol. Soc., London* **131**, 653–659.

Siever, R. *et al.* (1983). The dynamic Earth. *Sci. Am.* **249**, 30–144.

Silviera, D. J. (1981). "The Potential Influence of Organic Compounds on the Transport of Radionuclides from a Geologic Repository," Rep. PNL-3414. Pacific Northwest Lab., Richland, Washington.

Simmons, J. H. *et al.* (1979). Fixation of radioactive waste in high-silica glasses. *Nature (London)* **278**, 729–731.

SIPRI (1979). "Nuclear Energy and Nuclear Weapon Proliferation." Taylor & Francis, London.

Skinner, B. J. and Walker, C. A. (1982). Radioactive wastes. *Am. Sci.* **70**, 180–181.

Smedes, H. E. (1980). Rationale for geologic isolation of high-level radioactive waste and assessment of the suitability of crystalline rocks. *Geol. Surv. Open-File Rep. (U.S.)* **80–165**.

Smith, A. G. and Briden, J. C. (1977). "Mesozoic and Cenozoic Paleocontinental Maps." Cambridge Univ. Press, London and New York.

Smith, C. F. *et al.* (1980). "Hazard Index for Underground Toxic Material," Rep. UCRL-52889. Lawrence Livermore Lab., Livermore, California.

Smith, D. G., ed. (1982). "The Cambridge Encyclopedia of Earth Sciences." Cambridge Univ. Press, London and New York.

Smith, R. L. (1960). Ash flows. *Geol. Soc. Am. Bull.* **71**, 795–842.

Smith, R. M. (1981). "Radionuclide Distributions Around a Retired Nuclear Waste Disposal Well," Rep. RHO-SA-266. Rockwell Hanford Operations, Richland, Washington.

Smyth, J. R. *et al.* (1979). "A Preliminary Evaluation of the Radioactive Waste Isolation Potential of the Alluvium-filled Valleys of the Great Basin," Rep. LA-7962-MS. Los Alamos Natl. Lab., Los Alamos, New Mexico.

Sousselier, Y. (1981). Criteria for high-level waste disposal. *Radioact. Waste Manage.* 1, 359–364.

Spearing, D. R. (1971). Summary sheets of sedimentary deposits. *Geol. Soc. Am. Map and Chart Series MC-8.*

Spitsyn, V. I. *et al.* (1977). "Basic Requisites and the Practice of Using Deep Water Tables for Burying Liquid Radioactive Wastes," Transl. ORNL-tr-4390. Oak Ridge Natl. Lab., Oak Ridge, Tennessee.

Spoljaric, N. (1981). Geologic aspects of disposal of highly radioactive nuclear waste. *Open-File Rep.—Del. Geol. Surv.* 15, 1–45.

State Planning Council (1981). "Low-level Radioactive Waste Managment: Case Studies on Commercial Burial Sites." State Planning Council on Radioactive Waste Management, Washington, D.C.

Steinborn, T. L. *et al.* (1980). "Recommended New Criteria for the Selection of Nuclear Waste Repository Sites in Columbia River Basalt and U.S. Gulf Coast Domed Salt," Rep. UCRL-52749. Lawrence Livermore Lab., Livermore, California.

Stinton, D. P. *et al.* (1983). "Characterization of Hydrofracture Grouts for Radionuclide Migration," Rep. CONF-830451-4. Oak Ridge Natl. Lab., Oak Ridge, Tennessee.

Subseabed Disposal Program (1983). "The Subseabed Disposal Program: 1983 Status Report," Rep. SAND83-1387. Sandia Labs., Albuquerque, New Mexico.

Sugden, D. E. and John B. S. (1976). "Glaciers and Landscape." Arnold, London.

Sun, R. J. (1982). Selection and investigation of sites for the disposal of radioactive wastes in hydraulically induced subsurface fractures. *Geol. Surv. Prof. Pap. (U.S.)* 1215, 1–87.

Swedish Nuclear Fuel Supply Co. (1977). "Handling of Spent Nuclear Fuel and Final Storage of Vitrified High-level Reprocessing Waste," KBS-1. Swed. Nucl. Fuel Supply Co., Div. KBS, Stockholm.

Swedish Nuclear Fuel Supply Co. (1978). "Handling and Final Storage of Unreprocessed Spent Nuclear Fuel," KBS-2. Swed. Nucl. Fuel Supply Co., Div. KBS, Stockholm.

Swedish Nuclear Fuel Supply Co. (1983). "Final Storage of Spent Nuclear Fuel," KBS-3. Swed. Nucl. Fuel Supply Co., Div. KBS, Stockholm.

Talbot, C. J. and Jarvis, R. J. (1984). Age, budget, and dynamics of an active salt extrusion in Iran. *J. Struct. Geol.* 6, 521–533.

Tarandi, T. (1983). "Calculated Temperature Field in and around a Repository for Spent Nuclear Fuel," Tech. Rep. 83–22. Swed. Nucl. Fuel Supply Co., Div. KBS, Stockholm.

Tarling, D. H. ed. (1978). "Evolution of the Earth's Crust." Academic Press, London.

Templeton, W. L. (1980). "Dumping of Low-level Radioactive Waste in the Deep Ocean," Rep. PNL-SA-8840. Pacific Northwest Lab., Richland, Washington.

Templeton, W. L. and Preston, A. (1982). Ocean disposal of radioactive wastes. *Radioact. Waste Manage.* 3, 75–113.

Tera Corp. (1978). "Inventory and Sources of Transuranic Solid Waste," Rep. UCRL-13934. Lawrence Livermore Lab., Livermore, California.

Thoregren, U. (1983). "Final Disposal of Spent Nuclear Fuel - Standard Programme for Site Investigations," Tech. Rep. 83–31. Swed. Nucl. Fuel Supply Co., Div. KBS, Stockholm.

Thorpe, R. K. (1979). "Characterization of Discontinuities in the Stripa Granite - Time-Scale Experiment," Rep. LBL-7083. Lawrence Berkeley Lab., Berkeley, California.

Thorpe, R. K. and Springer, J. (1981). "Fracture Mapping for Radionuclide Migration Studies in the Climax Granite," Rep. UCID-19081. Lawrence Livermore Lab., Livermore, California.

Thunvik, R. and Braester, C. (1980). "Hydrothermal Conditions Around a Radioactive Waste Repository," Tech. Rep. 80–19. Swed. Nucl. Fuel Supply Co., Div. KBS, Stockholm.

Thury, M. (1980). "Erläuterungen zum NAGRA-Tiefbohrprogramm als vorbereitende Handlung im Hinblick auf das 'Projekt Gewähr'," Tech. Rep. 80–07. National Cooperative for Radioactive Waste Storage, Baden, Switzerland.

Tin Chan et al. (1978). "Theoretical Temperature Fields for the Stripa Heater Project," Rep. LBL-7082. Lawrence Berkeley Lab., Berkeley, California.

Tokunaga, T. and Narasimhan, T. N. (1982). "Recent Hydrological Observations from the Riverson and the Maybell Tailings Piles," Rep. LBL-15036. Lawrence Berkeley Lab., Berkeley, California.

Topp, S. V. (1982). "Scientific Basis for Nuclear Waste Management," Vol. IV. North-Holland, New York.

Trask, N. J. (1982). Performance assessments for radioactive waste repositories: The rate of movement on faults. Geol. Surv. Open-File Rep. (U.S.) 82-0972, 1–25.

Trimmer, D. et al. (1980). Effect of pressure and stress on water transport in intact and fractured gabbro and granite. JGR, J. Geophys. Res. 85, 7059–7071.

Trümpy, R. et al. (1980). "Geology of Switzerland." Wepf, Basel.

Turekian, K. K. (1976). "Oceans," 2nd ed. Prentice-Hall, Englewood Cliffs, New Jersey.

Turekian, K. K. and Rona, P. A. (1977). Eastern Atlantic fracture zones as potential disposal sites for radioactive waste. Environ. Geol. 2, 59–62.

Union of Concerned Scientists (UCS) (1975). "The Nuclear Fuel Cycle," 2nd ed. MIT Press, Cambridge, Massachusetts.

Upton, A. C. (1982). The biological effects of low-level ionizing radiation. Sci. Am. 246, 29–37.

U.S. Department of Energy (DOE) (1979a). "Progress Report on the Grand Junction Uranium Mill Tailings," Rep. DOE/EV-0033. USDOE, Washington, D.C.

U.S. Department of Energy (DOE) (1979b). "Uranium Mining, Milling and Conversion," Rep. DOE-EDP-58. USDOE, Washington, D.C.

U.S. Department of Energy (DOE) (1979c). "Report to the President by the Interagency Review Group on Nuclear Waste Management." USDOE, Washington, D.C.

U.S. Department of Energy (DOE) (1980). "Nevada Nuclear Waste Storage Investigations," Rep. NVO-196-16. Nevada Operations Office, Las Vegas, Nevada.

U.S. Department of Energy (DOE) (1982a). "Environmental Assessment. Waste Form Selection for SRP High-level Waste," Rep. DOE/EA-0179. USDOE, Washington, D.C.

U.S. Department of Energy (DOE) (1982b). "Spent Fuel and Radioactive Waste Inventories, Projections and Characteristics," Rep. DOE/NE-0017-1. USDOE, Washington, D.C.

U.S. Energy Research Development Administration (ERDA) (1975). "Final Environmental Impact Statement—Waste Management Operations, Hanford Reservation, Richland, Washington," Rep. ERDA-1538. USERDA, Washington, D.C.

U.S. Energy Research Development Administration (ERDA) (1976). "Alternatives for Managing Wastes from Reactors and Post-fission Operations in the LWR Fuel Cycle," Rep. ERDA-76-43. USERDA, Washington, D.C.

U.S. Energy Research Development Administration (ERDA) (1977). "Final Environmental Impact Statement - Waste Management Operations, Savannah River Plant, Aiken, South Carolina," Rep. ERDA-1537. USERDA, Washington, D.C.

van Olphen, H. and Veniale, F., eds. (1982). "International Clay Conference 1981." Elsevier, Amsterdam.

Verkerk, B. (1981). Comparison of long-term release consequences of spent fuel and vitrified waste repositories in a salt formation. Radioact. Waste Manage. 1, 337–357.

Vogler, S. et al. (1983). "Materials Recycle and Waste Management in Fusion Power Reactors," Rep. ANL/FPP/TM-163. Argonne Natl. Lab., Argonne, Illinois.

Vorren, T. O. (1979). Weichselian ice movements, sediments and stratigraphy on Hardangervidda, South Norway. Nor. Geol. Unders. 350, 1–117.

Waddell, R. K. (1982). Two-dimensional, steady-state model of ground-water flow, Nevada Test Site and vicinity, Nevada-California. Water Resour. Invest. (U.S. Geol. Surv.) 82-4085, 1–72.

Wahi, K. K. et al. (1978). "Thermomechanical Response of a One-level Spent Fuel Nuclear Waste Repository," Rep. Y/OWI/SUB-78/16549/1. Office of Waste Isolation, Oak Ridge, Tennessee.

Wahi, K. K. et al. (1980). "Numerical Simulations of Earthquake Effects on Tunnels for Generic Nuclear Waste Repositories," Rep. DP-1579. Savannah River Lab. Aiken, South Carolina.

Waller, R. M. et al. (1978). Potential effects of deep-well waste disposal in western New York. Geol. Surv. Prof. Pap. (U.S.) 1053, 1–39.

Walter, G. R. (1982). "Theoretical and Experimental Determination of Matrix Diffusion and Related Solute Transport Properties of Fractured Tuffs from the Nevada Test Site," Rep. LA-9471-MS. Los Alamos Natl. Lab., Los Alamos, New Mexico.

Wang, H. F. and Anderson, M. P. (1982). "Introduction to Groundwater Modeling." Freeman, San Francisco, California.

Wang, J. S. Y. et al. (1981). A study of regional temperature and thermohydrologic effects of an underground repository for nuclear wastes in hard rock. JGR, J. Geophys. Res. 86, 3759–3770.

Wang, J. S. Y. et al. (1983). The state-of-the-art of numerical modeling of thermohydrologic flow in fractured rock masses. Environ. Geol. 4, 133–199.

Wanless, R. W. et al. (1960). Sulphur isotope investigation of the gold–quartz deposits of the Yellowknife district. Econ. Geol. 55, 1591–1621.

Washburn, H. L. (1967). Instrumental observation of mass wasting in the Mesters-Vig district, northeast Greenland. Medd. Groenl. 166, 1–296.

Washburn, J. F. et al. (1980). "Multicomponent Mass Transport Model: A Model for Simulating Migration of Radionuclides in Groundwater," Rep. PNL-3179. Pacific Northwest Lab., Richland, California.

Weart, W. D. (1981). Technical developments in the W.I.P.P. program. Bull. Assoc. Eng. Geol. 18, 423–436.

Webb, G. A. M. (1980). The tripod on which the safety assessment of disposal rests. Radioact. Waste Manage. 1, 113–119.

Weber, W. J. and Roberts, F. P. (1983). A review of radiation effects in solid nuclear waste forms. Nucl. Technol. 60, 178–198.

West, J. M. et al. (1982). Microbes in deep geological systems and their possible influence on radioactive waste disposal. Radioact. Waste Manage. 3, 1–15.

Westbrook, G. K. (1982). The Barbados Ridge Complex: tectonics of a mature forearc system. Geol. Soc. London Spec. Publ. 10, 275–290.

Westsik, J. H. and Peters, R. D. (1980). "Time and Temperature Dependence of the Leaching of a Simulated High-level Waste Glass," Rep. PNL-SA-9054. Pacific Northwest Lab., Richland, Washington.

Westsik, J. H. and Turcotte, R. P. (1978). "Hydrothermal Reactions of Nuclear Waste Solids. A Preliminary Study," Rep. PNL-2759. Pacific Northwest Lab., Richland, Washington.

White, A. F. and Delany, J. M. (1982). "Chemical Transport Beneath a Uranium Mill Tailings Pile, Riverton, Wyoming," Rep. LBL-15038. Lawrence Berkeley Lab., Berkeley, California.

Wikberg, P. *et al.* (1983). "Redox Conditions in Groundwaters from Svartoberget, Gideå, Fjällveden and Kamlunge," Tech. Rep. 83–40. Swed. Nucl. Fuel Supply Co., Div. KBS, Stockholm.

Wilder, D. G., and Patrick, W. C. (1981). Geotechnical status report for test storage of spent reactor fuel in Climax Granite, Nevada Test Site. *Bull. Assoc. Eng. Geol.* **18**, 355–367.

Winograd, I. J. (1981). Radioactive waste disposal in thick unsaturated zones. *Science* **212**, 1457–1464.

Winograd, I. J. and Thordarson, W. (1975). Hydrogeologic and hydrogeochemical framework, South-Central Great Basin, Nevada-California, with special reference to the Nevada Test Site. *Geol. Surv. Prof. Pap. (U.S.)* **712-C**, 1–126.

Witherspoon, P. A. (1980). "Radioactive Waste Storage in Mined Caverns in Crystalline Rock - Results of field Investigations at Stripa, Sweden," Rep. LBL-11651. Lawrence Berkeley Lab., Berkeley, California.

Witherspoon, P. A. *et al.* (1981). Geologic storage of radioactive waste: field studies in Sweden. *Science* **211**, 894–901.

Wittke, W., ed. (1982). "Rock Mechanics: Caverns and Pressure Shafts." Balkema, Rotterdam.

Wood, B. J. and Walter, J. V. (1983). Rates of hydrothermal reactions. *Science* **222**, 413–415.

Wyllie, P. J. (1980). The origin of kimberlite. *JGR, J. Geophys. Res.* **85**, 6902–6910.

Wynne, B. (1982). "Rationality and Ritual: The Windscale Inquiry and Nuclear Decisions in Britain." British Society of the History of Science, London.

Wyss, M. (1979). Estimating maximum expectable magnitude of earthquakes from fault dimensions. *Geology* **7**, 336–340.

Young, A. (1969). Present rate of land erosion. *Nature (London)* **224**, 851–852.

Zehner, H. H. (1983). Hydrogeologic investigation of the Maxey Flats radioactive waste burial site, Fleming County, Kentucky. *Geol. Surv. Open-File Rep. (U.S.)* **83-133**, 1–148.

Zeller, E. J. and Saunders, D. F. (1972). "Radioactive Waste disposal. A Suggestion for a Permanent International Polar High-level Radioactive Waste Repository," Rep. COO-1057-59. Space Technol. Lab., Univerisity of Kansas, Lawrence.

Zellmer, J. T. (1981). "Stability of Multilayer Earthen Barriers used to Isolate Mill Tailings: Geologic and Geotechnical Considerations," Rep. PNL-3902. Pacific Northwest Lab., Richland, Washington.

Zoback, H. D. and Anderson, R. N. (1982). Ultrasonic borehole televiewer investigation of ocean crustal layer 2A, Costa Rica Rift. *Nature (London)* **295**, 375–379.

INDEX